CHAPTER 1

INTRODUCTION

Romeo M. Flores

U.S. Geological Survey

Denver, Colorado 80225

The understanding of fluvial environments and processes that operate within them as well as their products in the geological record is a recent development. During the past decade, facies analysis of fluvial rocks has been increasingly used to explore for and develop hydrocarbons, coal, uranium, and metallic minerals. The rapid growth in the data base on fluvial depositional systems coupled with the need to recognize the economic potential of their deposits has yielded numerous excellent books (Miall, 1978; Collinson, 1978; Galloway and Hobday, 1983; Collinson and Lewin, 1984), which deal with recognition and classification of the whole spectrum of fluvial systems, fluvial processes and their products, facies models of ancient fluvial deposits, and application of fluvial models to resource exploration and development. This notebook is an outgrowth of the burgeoning geological investigations of fluvial rocks and their associated resource potential.

An overall view of the elements of deposition, framework, and styles of aggradation of fluvial systems is the key to comprehending fluvial architecture. Recognition of the depositional elements and aggradational patterns provides a classification of fluvial systems (e.g., alluvial fans and braided, meandering, anastomosed, and basin trunk-tributary rivers). These fluvial systems frequently are integrated and interconnected drainage networks. Factors that control the development of a specific type of fluvial system include tectonic, climatic, hydraulic, and geomorphic conditions. Processes and products of a fluvial system are influenced by these factors.

A synthesis of the processes and products of fluvial systems is best derived from recent fluvial systems, with emphasis on physical characteristics that occur in each system. Identification of the variations in properties of various types of fluvial sediments defines facies characteristics that are used to develop facies models of ancient deposits. Analysis of facies models of ancient deposits focuses on variations of facies sequences, relationships, and associations. These facies characteristics are best described by facies stratigraphy, which permits an understanding of the vertical and lateral variations of facies leading to construction of predictive models that can be used in exploration and development of energy and mineral resources associated with fluvial deposits. The economic applications of fluvial models to exploration and development will be developed from surface and subsurface data.

The notebook is divided into 11 chapters that cover methodology and classification of fluvial systems as well as modern and ancient deposits of alluvial fans, fan deltas, braided streams, meandering streams, and anastomosed streams. In addition, application of facies modeling to exploration and development of hydrocarbons, coal, and uranium is discussed for the Rocky Mountain region, Mid Continent, Gulf Coast, and western China.

In Chapter 2, Ethridge summarizes surface and subsurface methods of investigating and classifying fluvial systems. He notes that the inference of depositional environments from ancient sedimentary rocks can be derived from vertical sequences that occur in time and space. These vertical sequences can be described according to their mineral composition and texture, sedimentary structures, architecture, and sedimentary sequences and models. Vertical sequences are particularly useful in subsurface investigations of fluvial rocks. Data from subsurface investigations include drill cuttings, continuous

RECOGNITION OF FLUVIAL DEPOSITIONAL SYSTEMS AND THEIR RESOURCE POTENTIAL

Lecture Notes for Short Course No. 19
Sponsored by the Society of Economic
Paleontologists and Mineralogists

Romeo M. Flores, U.S.G.S., Denver
Frank G. Ethridge, Colo. State Univ., Fort Collins
Andrew D. Miall, Univ. of Toronto, Toronto
William E. Galloway, Bur. of Econ. Geol., Univ. of Texas, Austin
Thomas D. Fouch, U.S.G.S., Denver

Printed in U.S.A.
1985

CONTENTS

cores, wireline logs, and other seismic records. Log patterns may produce data for interpreting depositional environments. These environments may be classified in areal view, especially fluvial channels, into braided, meandering, and anastomosed channel types.

In Chapter 3, Miall proposes a new method of facies analysis of fluvial rocks using architectural-element analysis. He suggests that vertical profiles and channel morphology no longer serve as diagnostic tools because these features can be produced by various processes and geomorphic settings. In addition, the morphology of different channel types grades into one another and the internal elements (e.g., lateral accretion, bars) occur in many channel types. This approach to facies modeling of fluvial deposits involves analysis of eight basic architectural elements that include: 1) element CH: channels; 2) element GB: gravelly bars and bedforms; 3) element SB: sandy bedforms; 4) element FM: foreset macroforms; 5) element LA: lateral accretion deposits; 6) element SG: sediment gravity flow deposits; 7) element LS: laminated sand sheets; and 8) element OF: overbank fines. Twelve fluvial models can be developed based on the sinuosity, braiding parameter, sediment type, and the eight characteristic elements. These models are purely descriptive and geomorphic, which free users from misleading assumptions.

In Chapter 4, Miall uses architectural-element analysis to classify multiple-channel bedload rivers. The bedload rivers are commonly known as gravel- or sand-rich sediment load rivers or braided rivers that imply low sinuosity streams; however, higher sinuosities are common. His classification comprises six architectural models--3, 4, 9, 10, 11, and 12. Model 3 corresponds to the Donjek River, Yukon in Alaska and model 4 is exemplified by the Endrick River in Scotland, by the Nueces River in Texas, and by the lower Babbage River, Yukon in Alaska. Model 9 is represented by the Platte River and model 10 is typified by the south Saskatchewan River. Model 11 resembles sand-bed ephemeral streams of central Australia and model 12 is similar to the Bijou Creek, Colorado.

Modern alluvial fans and fan deltas are discussed by Ethridge in Chapter 5. Alluvial fans are cone-shaped deposits built by streams from highlands to adjoining lowlands in arid and humid regions. Deposits of alluvial fans are characterized by coarse grained, poorly sorted gravels and sands. Grain size and thickness of these deposits decrease downfan. Particle roundness increases downfan. Vertical grain size trends and sequences are controlled by basin-margin faults. Fan deltas consist of alluvial fans that build into an ocean, sea or lake. Fan-delta deposits are coarsening-upward sequences that contains gravel and coarse sand-sized sediments. Three depositional fan delta models include shelf, slope, and Gilbert-type.

Galloway discusses ancient deposits of alluvial fans and fan deltas in Chapter 6. Alluvial fans are recognized by their lobate morphology, radial sediment dispersal pattern, dominantly transverse gravel to sand composition, ground-water flow, relationships to nearby structural uplands, and association with lacustrine or through-flowing streams. Fan deltas are alluvial fans that debouch directly into a standing body of water (e.g., lake, ocean). Examples that illustrate the properties of ancient alluvial fan and fan delta deposits are selected from Pennsylvanian rocks (Atokan to Desmoinesian) in New Mexico, Triassic rocks (Hawkesbury Sandstone) in the Sydney Basin, Australia, and Neogene rocks (Ogallala Formation) in Texas. These alluvial fans are dominated by braided-stream processes that locally grade into mixed-load meanderbelt tracts separated by lacustrine floodbasins. The characteristics of these alluvial fans and their deposits are comparable to those of the Kosi River fan in Nepal and India.

Modern and ancient deposits of meandering streams are described by Galloway in Chapter 7. His classification of meandering streams followed Schumm's (1981) classification based on: 1) the ratio of sediment load to suspended load transported by a stream; 2) cross-sectional geometry of the channel expressed as width/depth ratio; and 3) independent of slope, discharge, and periodicity of flow. These relationships are quantified in modern streams providing quantitative trends that can be applied in the classification of ancient fluvial systems. This classification scheme yields bedload, mixed-load, and suspended-load types of fluvial channels. Bedload channels represent braided and coarse-grained meandering streams. Mixed-load channels reflect meandering streams with channel plug deposits. Suspended-load channels represent anastomosed streams encased in fine-grained floodbasin deposits.

Flores discusses coal deposits in Cretaceous and Tertiary fluvial systems of the Rocky Mountain region in Chapter 8. Alluvial plain environments landward of Late Cretaceous maximum transgressive coastlines and Late Cretaceous-early Tertiary regressive coastlines, and in Tertiary intermontane settings provide a varied milieu of coal accumulation in the Rocky Mountain region. In the Gallup sag, New Mexico, coals formed in swamps associated with a lower alluvial plain that formed landward of coastal barriers. The swamps were sustained by regional transgression of coastal lagoons, which caused a rise of the ground-water table. In the Raton Basin, Colorado and New Mexico, coals formed in swamps on the upper alluvial plain. Here, the alluvial plain and coal-forming environments are controlled by extrabasinal and intrabasinal tectonism, which produced megarhythmic sequences of coaly and sand-rich deposits. In the Powder River Basin, the accumulation of coal in swamps of the intermontane alluvial plain is controlled by autocyclic and allocyclic processes. The coal beds in this intermontane setting are anomalously thick and laterally extensive. In the Raton and Powder River Basins, upward motion of deep-basin ground water probably sustained the swamps on the alluvial plain.

The investigation of the alluvial plain facies in the Gallup sag, Raton Basin, and Powder River Basin provides a key to understanding fluvial facies related to coal deposition and shallow gas accumulation. Mapping the distribution and geometry of coal deposits (source facies) is recommended as an alternative to predicting, recognizing, and delineating reservoir facies such as channel and crevasse splay sandstones in coal-bearing rocks. Coal is easily identified in geophysical logs and its signature is not adversely affected by diagenesis. Thus, coal can be a useful tool in facies-oriented exploration and development of hydrocarbons.

In Chapter 9, Ethridge reviews the reservoir characteristics of ancient fluvial deposits, primarily in the Rocky Mountain and Mid Continent region. Reservoir units consist of: 1) point-bar sandstones of meandering streams, which are sealed mainly by abandoned channel plugs or fine-grained floodplain deposits; 2) gravelly channel sandstones of braided streams, which are more homogeneous than the point-bar reservoirs; and 3) fan-delta deposits. These reservoir rocks occur either as stratigraphic, combined stratigraphic-structural, or subunconformity traps.

In Chapter 10, Fouch discusses oil and gas-bearing Upper Cretaceous and Paleocene fluvial rocks in central and northeast Utah. These rocks are the major reservoir units for large volumes of hydrocarbons in Upper Cretaceous and Paleogene rocks. The hydrocarbons are in tight or unconventional reservoir sandstones with well developed secondary porosity. In Paleogene rocks, these reservoir units were deposited in alluvial, marginal lacustrine, and open lacustrine environments. The alluvial facies include deposits of alluvial fan and high mud-flat environments. The marginal lacustrine facies consists of deltaic and interdeltaic deposits. The Upper Cretaceous reservoir sandstones were deposited in alluvial fans, braided streams, and meandering streams. The principal reservoir rocks deposited by braided streams are blanketlike conglomeratic channel sandstones exemplified by the oil and gas bearing rocks in the Tavaputs Plateau. Reservoir units formed by meandering streams consist of lenticular to sheetlike channel sandstones represented by the gas-bearing Tuscher, Farrer, and Neslen Formations.

The reservoir units in Paleogene rocks vary from conglomeratic rocks of alluvial fans to single and composite, lenticular channel sandstones of fluvial and deltaic origin. Carbonates with little or no cement, which formed in marginal-lacustrine environments, also served as reservoir units. The open-lacustrine rocks were the principal source beds for oil accumulation in Paleogene rocks. Oil and gas pools in Upper Cretaceous and Tertiary rocks are considered to occur in stratigraphic traps. Porosity in reservoir rocks are generally secondary, a phenomenon that indicates that the traps should be called diagenetic stratigraphic traps. Similar fluvial reservoir beds have been explored in China which contain much larger reserves than those in Utah or other parts of the United States.

Hydrocarbons in fluvial deposits of the Gulf Coast region are discussed by Galloway in Chapter 11. Deposits of bedload, mixed load, and suspended load fluvial systems are host to prolific hydrocarbon accumulation in Tertiary deposits (Frio/Catahoula Formations) in the Gulf coast region. Petroleum reservoir units consist of multiple channel-fill and crevasse splay sandstones encased in mud-rich floodplain sequences.

REFERENCES CITED

Collinson, J. D., 1978, Alluvial sediments, in Reading, H. G., ed., Sedimentary Environments and Facies: Elsevier, N. Y., p. 15-59.

Collinson, J. D., and Lewin, J., (eds.), 1984, Modern and ancient fluvial systems: Inter. Assoc. Sediment., Special Publications 6, 575 p.

Galloway, W. E., and Hobday, D. K., 1983, Terrigenous Clastic Depositional Systems: Springer-Verlag, N. Y., 423 p.

Miall, A. O., (ed.) 1978, Fluvial Sedimentology: Can. Soc. Petrol. Geol., Mem. 5, 859 p.

Schumm, S. A., 1981, Evolution and response of the fluvial system, sedimentological implications, in Ethridge, F. G., and Flores, R. M., eds., Recent and Ancient Nonmarine Depositional Environments: Models for Exploration: Soc. Econ. Paleon. Mineral., Spec. Pub. 31, p. 19-30.

CHAPTER 2

SURFACE AND SUBSURFACE METHODS OF INVESTIGATION AND CLASSIFICATION OF FLUVIAL SYSTEMS

Frank G. Ethridge
Department of Earth Resources
Colorado State University

Objectives and Definitions

The objectives for the first day of the course are to review sediment processes that result in erosion, transportion and deposition in modern fluvial systems, and to present criteria and models that are used to reconstruct ancient fluvial and related deposits. The criteria and models were developed, for the most part, from studies of modern depositional environments. Although environments can be erosional, equilibrium or depositional in nature, it is the Depositional Environment that will be our principal concern. The depositional environment is preserved in the rock record as a three-dimensional body with framework (usually sandstones and conglomerates) and non-framework (usually mudstones) deposits. Assemblages of related depositional environments are referred to as Depositional Systems.

Another equally important objective is to apply the knowledge of modern and ancient environments to the practical problems of exploring for an exploiting mineral and fossil fuel resources that may be contained within ancient fluvial deposits. This aspect of the course will be covered on the second day.

Criteria for Reconstructing Ancient Environment

General

Reconstruction of ancient depositional environments in stratigraphic sequences is based on two fundamental concepts. The first, Johannes Walther's Law of Facies (1894) states: "a conformable vertical sequence of facies was generated by a lateral sequence of environments" (Selley, 1976). The second concept, stated by Visher (1965) and formalized by Selley (1976)specifies that there are a finite number of sedimentary environments and processes on the earth's surface and a finite number of idealized vertical sequences that occur in time and space in the geologic record. If recognized these idealized sequences can be used to infer depositional environments of ancient sedimentary sequences.

To recognize the sequences we must first understand the fundamental and derived properties of sedimentary rocks. Each property's usefulness will be reviewed briefly before attempting to describe the various fluvial environments.

Composition and Texture

One of the earliest methods used to distinguish depositional environments was texture. It is generally accepted that grain size and shape are environmentally sensitive. Shape, however, is difficult to measure in sand size sediments, and the nature of the grain size-environment interaction is poorly understood despite a plethora of publications on the subject.

The relationship between size and composition provides an environmentally sensitive indicator in portions of a basin with a single source (provenance; Davies and Ethridge, 1975). Furthermore, size and composition data can be obtained from very small samples such as oil field cuttings. The value of textural and compositional petrographic data in oil and gas exploration has been illustrated by a study of the Muddy Formation in the northeast Powder River Basin (Davies, 1976). Good outcrop and core control in the area make it possible to recognize fluvio-deltaic systems, barrier bars and delta destructional bars. Figure 1 illustrates that for a given grain size, bar sandstones are more quartzose than fluvio-deltaic sandstones. Petrographic analysis of small samples under certain conditions can provide data to help differentiate environments and thus the possible sandstone reservoirs trends. However, the technique must be applied carefully because of the effect on sediment composition of tectonic framework, source, transport, diagenesis, size and environment.

Sedimentary Structures

The most useful criteria for interpreting ancient depositional environments are primary sedimentary structures. Specific structures can be used to reconstruct bed configurations and the hydrodynamic conditions at the time of deposition and vertical sequences of structures can be used to interpret specific depositional environments. The characteristics of primary sedimentary structures and their hydrodynamic interpretations are reviewed by Allen (1970), Collinson and Thompson (1982), Harms, Southard and Walker (1982) and in numerous publications and books. Consequently only a few brief remarks will be made here concerning the more common types of structures found in fluvial deposits.

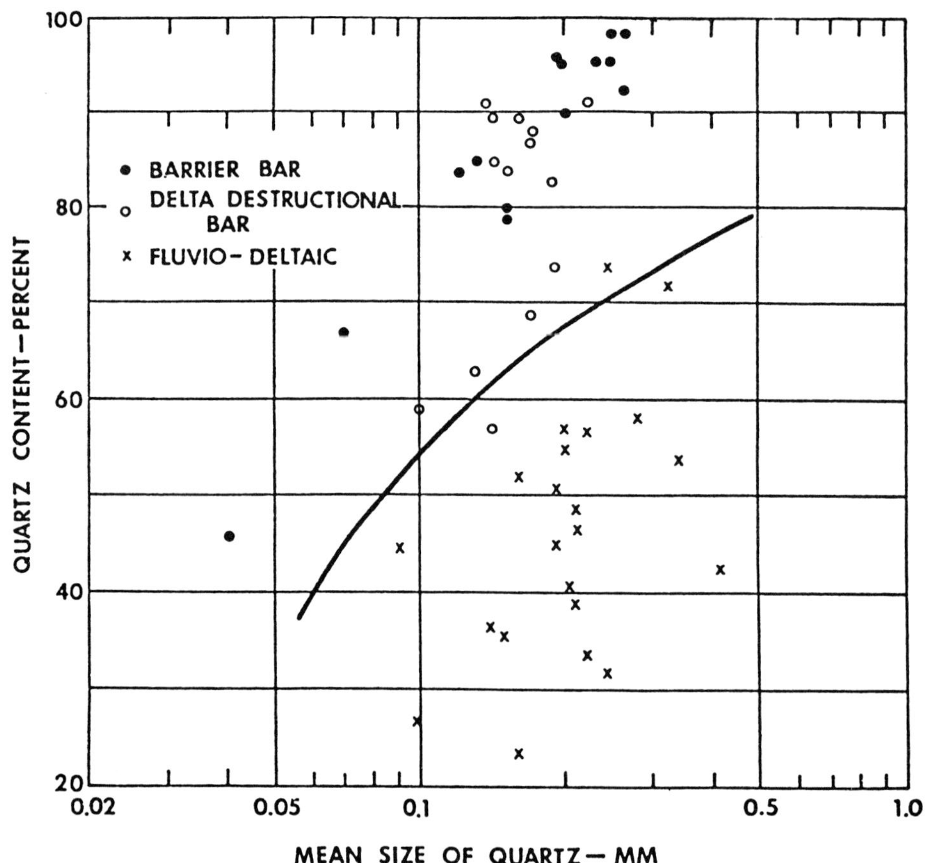

Figure 1. Quartz mean size versus content for Muddy Sandstone from constrasting environments, northeast Powder River Basin, Montana and Wyoming (from Davies, 1976).

Ripples form when the velocity of water flowing over an unconsolidated sand bed exceeds a certain critical value (which varies according to depth, grain size, etc.). The earliest ripples are straight crested, but with increased water velocity the ripple form changes to sinuous, and then to a more three-dimensional linguoid form (Fig. 2A). Bed aggradation under these conditions results in various types of small-scale, internal cross-stratification (Fig. 2B). Some examples of small-scale cross-stratification referred to as ripple drift

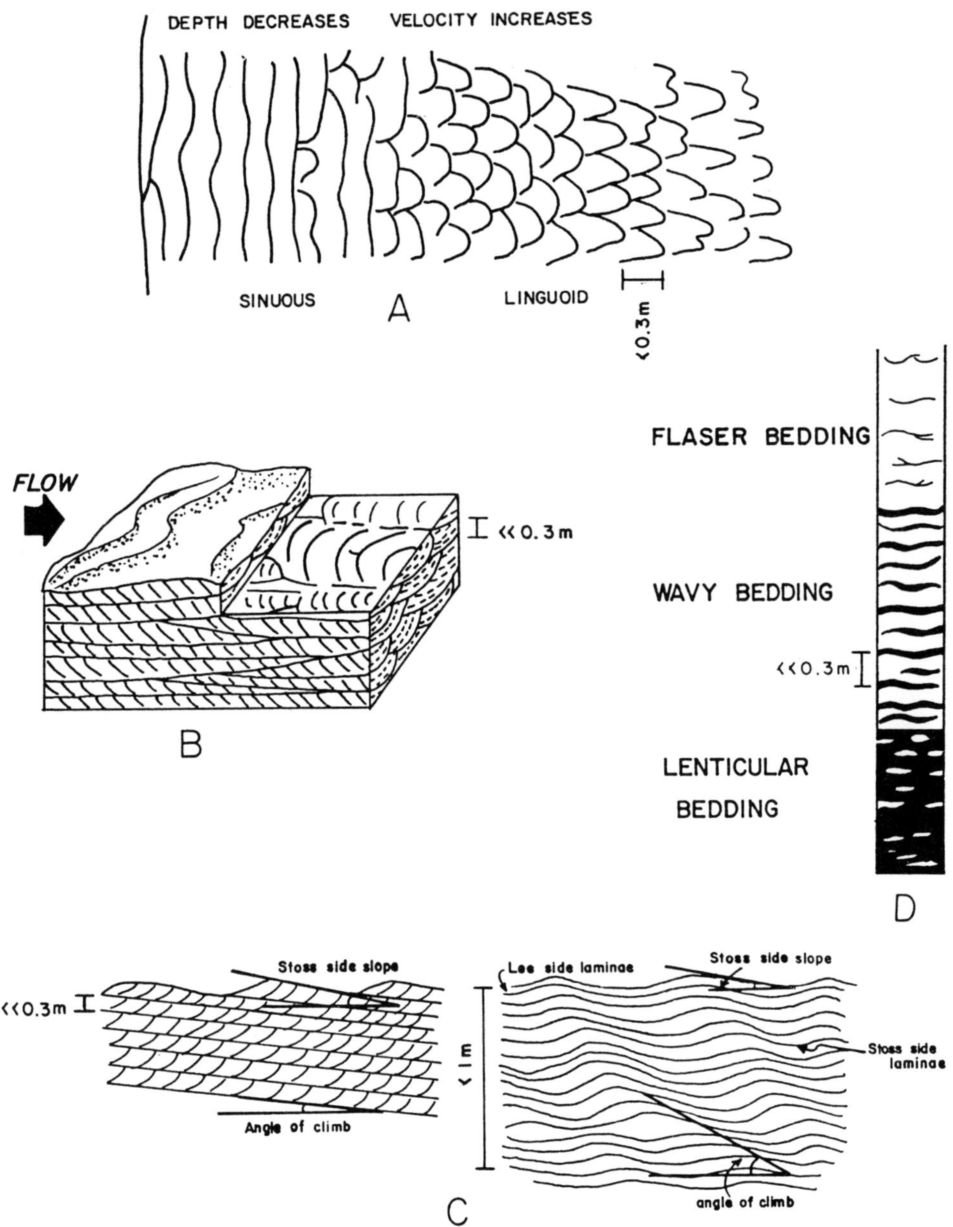

Figure 2. (A) Plan view morphology of ripples with increasing depth and velocity. (B) Small-scale cross lamination associated with unidirectional current ripples. (C) Schematics of different types of climbing ripples (ripple drift) cross stratification related to angle of climb and the angle of the stoss silde slope. (D) Variety of cross laminations resulting from mixed sand and mud lithologies (modified from Allen, 1968; Reinick and Collinson and Thompson, 1982).

or climbing ripple display roughly horizontal or inclined boundaries between sets. The angle of these bounding surfaces is determined by the relative rate of vertical aggradation compared to downstream (horizontal) migration. In these structures, set boundaries dip in opposite directions to the dip of the foreset laminations and may even be non-erosive and preserve stoss side laminations if aggradation is occurring very fast (Fig. 2C). In other units of ripple cross-stratified sand the structure is interrupted by laminations of mud or silt. If the sand occurs as isolated bodies of ripple forms, the structure is referred to as lenticular bedding (Fig. 2D). If the mud forms a minor component of the structure and drapes the ripple forms or only the troughs of ripples the structure is referred to as flaser bedding (Fig. 2D). Ripples and ripple cross-stratification are commonly preserved in fluvial deposits formed in fine-grained meanderbelt and ephemeral braided stream systems.

Large-scale primary structures include planar and trough cross-stratification. Planar or tabular cross-stratification (Fig. 3A) is characterized by subparallel set boundaries, the lack of strong conformity between foreset laminae and lower set boundaries in views perpendicular to flow, and straight to slightly sinuous laminae traces as viewed on horizontal surfaces. Planar cross-strata are produced by the aggradation and migration of sand waves or any straight migrating slipface (2-dimensional, large-scale ripples; Harms, et al., 1982). Under unidirectional flow within the same channel planar cross-stratification indicates flow velocities between those which produce ripples and those that produce dunes. Trough cross-stratification consists of erosional scours filled with curved

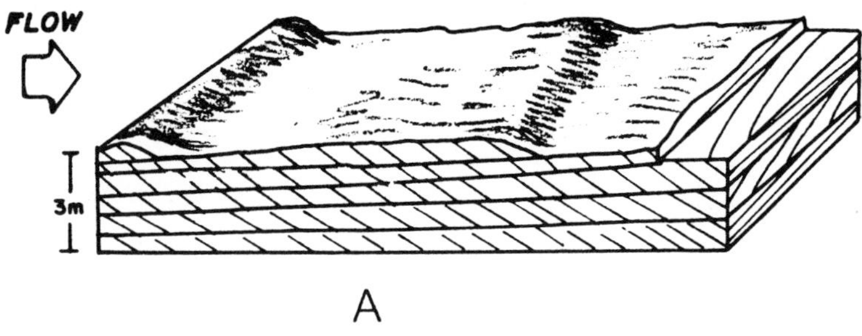

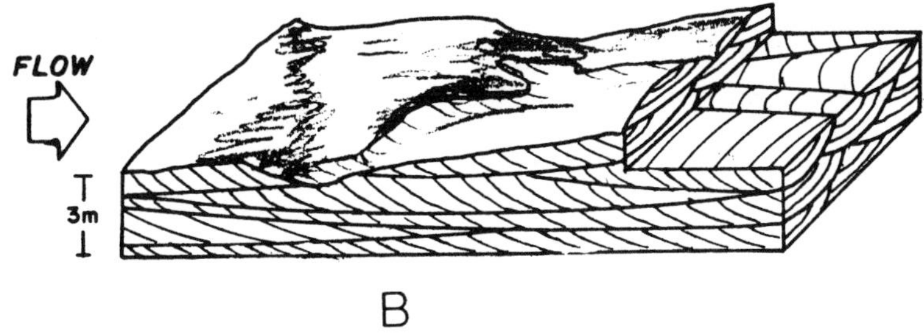

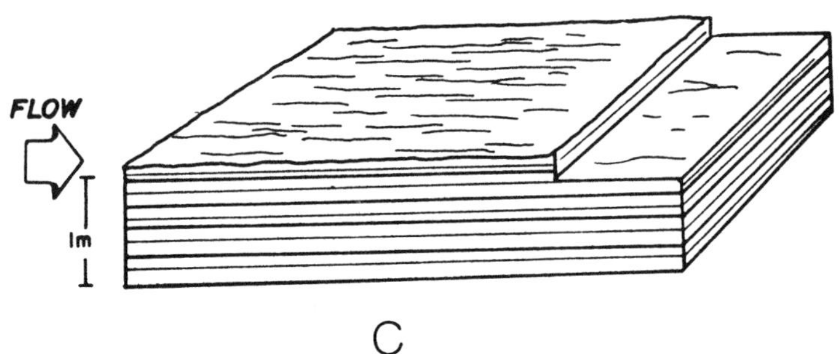

Figure 3. Cross bedding patterns resulting from migration of two-dimensional, large-scale ripples (A); three-dimensional, large-scale ripples, (B); and aggradation of plane bed under upper flow regime conditions (modified from Harms, et al., 1982).

foreset laminae (Fig. 3B). Trough cross-strata are produced by the migration and aggradation of dune bedforms (3-dimensional, large scale ripples; Harms, et al., 1982). Horizontal stratification (Fig. 3C) is defined as tabular sets of horizontal or near horizontal laminae in coarse silt, sand or gravel. Stratification develops from bedload deposition of material on a plane bed under higher flow velocities than those that normally produce dune forms. Bedding surfaces may show parting (current) lineations that are oriented parallel to flow direction. Parallel stratification can occur in fine-grained silts and muds and in coarser sediments which lack parting lineation. Fallout from suspension rather than transport of bedload material under high flow conditions forms parallel stratification. Planar and trough cross-stratification and horizontal and parallel stratification are common in many different types of braided, meandering and anastomosing fluvial systems.

Paleocurrent patterns derived from sedimentary structures are especially useful in interpreting facies. Structures such as channels and and groove marks yield only a sense of current flow while cross-stratification and pebble imbrication indicate both a sense of flow and a direction (Selley, 1976). Paleocurrent data are commonly presented as rose diagrams (Fig. 4). Fluvial deposits usually display a unimodal pattern with varying degrees of dispersion. Allen (1966) and Miall (1977) discuss problems that may be encountered in the collection, manipulation and interpretation of paleocurrent data.

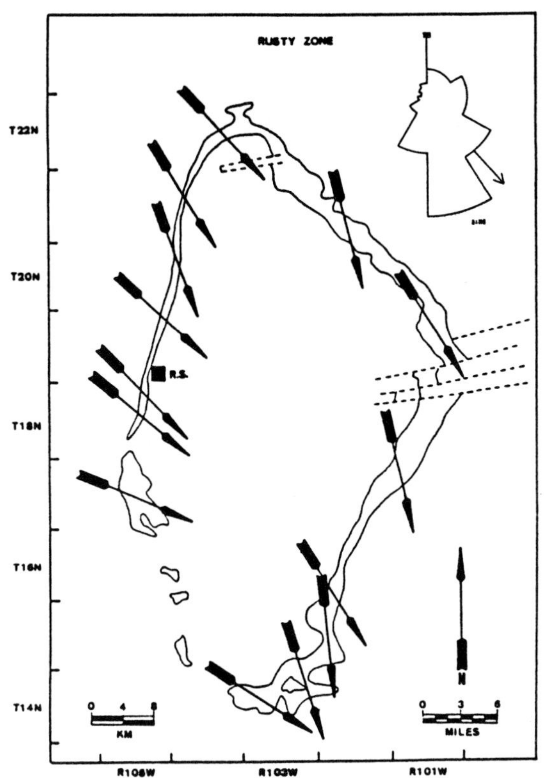

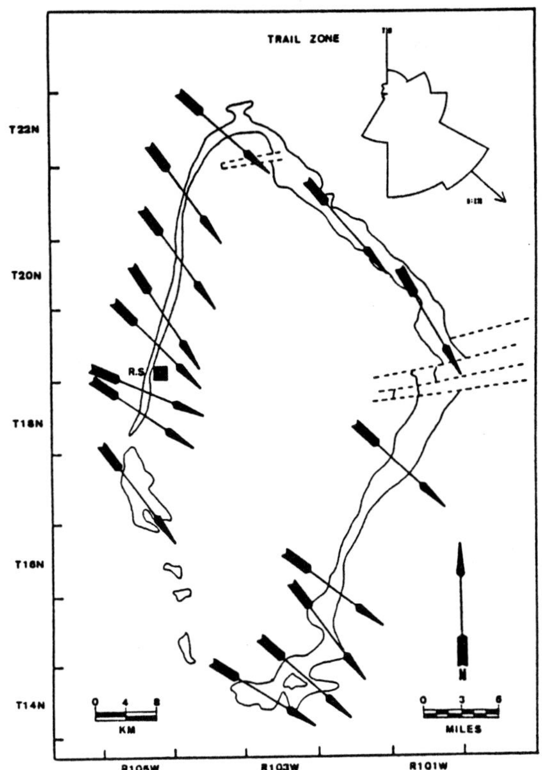

Figure 4. Mean paleocurrent vectors and paleocurrent rose diagrams of cross beds for two different zones of the fluvial Ericson Sandstone near Rock springs, Wyoming (after Sherman, 1983). Streams that deposited sediments of the Trail Zone had a higher sinuosity.

Architecture

The architecture of a sedimentary sequence includes the geometry and the internal arrangement of framework (e.g., channel) and non-framework (e.g., overbank) deposits. The internal arrangement and interconnectedness of sandstone units are important in determining reservoir/ore body continuity and in the analysis of modern and ancient groundwater systems (Miall, 1977). A new method of architectural-element analysis applied specifically to fluvial deposits is reviewed by Miall in the next chapter. Generally, fluvial sandstone body types can be classified into three principal sandstone geometry models (Fig. 5A; Harris and Hewitt, 1977) or into three cross-sectional geometries based on width/thickness ratios (Fig. 5B; Casey, 1980 and Friend, et al., 1979). Not all fluvial sandstone bodies fit into the scheme, but, as a first approximation of geometry, they prove useful. Problems associated with geometry prediction and sandstone body distribution are reviewed by Rittenhouse (1961) and Potter (1967). Brown (1969) discusses the role sediment compaction plays in distorting original geometry and arrangment of fluvial and deltaic sandstones in cratonic settings.

Sedimentary Sequences and Models

Visher (1965) was probably one of the first to formalize the use of sedimentary sequences in the interpretation of depositional environments. He produced a series of six unique sequences based on vertical succession of grain size, sorting, lithology, sedimentary structures, and geometry. Other authors have since amplified and expanded the original sequences and have developed models for each of the major depositional environments and systems. Summaries of the models have been published by Reading (1978), David (1983), Galloway and Hobday (1983) and Scholle and Spearing (1982). One example of a vertical sequence model for a type of fan-delta system is shown in Figure 6. This model will be discussed in Chapter 5.

In the following chapter Miall discusses the limitations imposed by using single vertical profile models and proposes a new method of analysis for fluvial deposits that involves subdividing the deposits into local suites of one or more of a set of eight basic three-dimensional, architectural elements. For correct diagnosis, however, the new approach is limited to outcrop exposures of at least

MAPS

BELT CONTINUOUS SHEET DISCONTINUOUS SHEET

CROSS SECTIONS

VERTICAL STACKING LATERAL STACKING ISOLATED STACKING

A

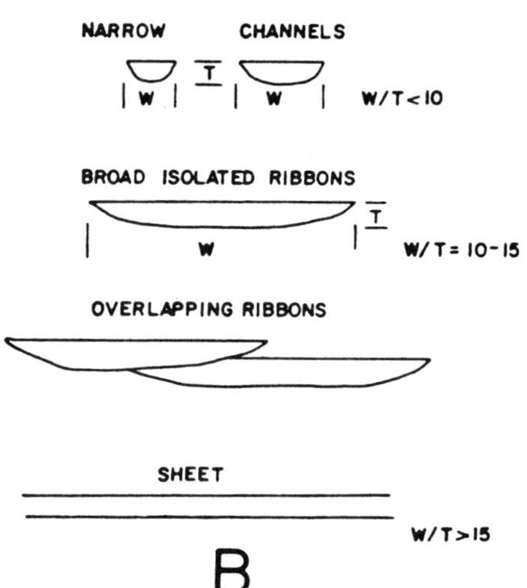

Figure 5. (A) Principal types of sandstone reservoir geometries illustrated for fluvial sandstone bodies (modified from Harris and Hewitt, 1977). (B) Cross-sectional geometries of fluvial sandstone and conglomerate bodies (modified from Casey, 1980; terminology modified after Friend, et al., 1979).

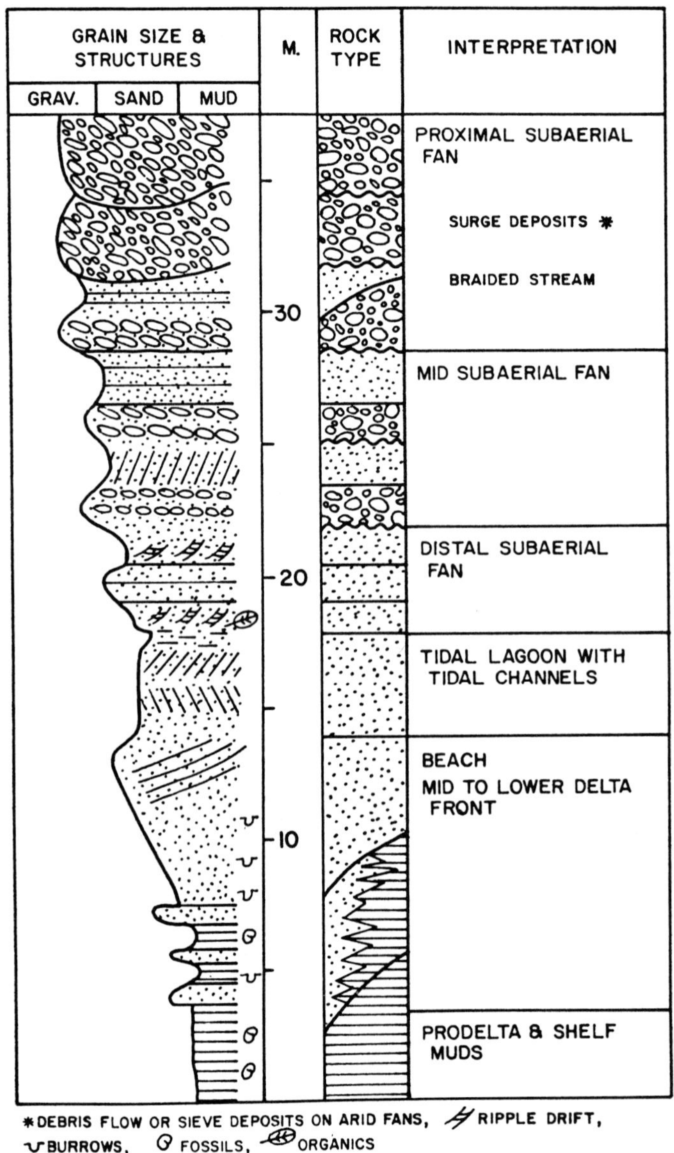

Figure 6. Vertical sequence model of hypothetical shelf-type fan delta deposits (from Ethridge and Wescott, 1984); See Chapter 5 for discussion and references.

tens of meters across and with a certain amount of three-dimensional control. The traditional vertical profile models will, therefore, find continued use, especially in subsurface investigations.

Subsurface Methods of Investigation

General

Available data sources for subsurface investigations include cuttings, continuous cores, wireline logs and seismic records. In this course, we will emphasize using continuous cores and wireline logs to obtain information for iterpreting fluvial and related depositional models and for basin analysis.

Lithologic Logs of Cores

Careful examination and recording of information from continuous cores can provide excellent data for environmental interpretation. Cores even have some advantages over conventional outcrop exposures. These advantages, summarized by Weimer and Tillman (1980) include: (1) not limited to a stratigraphic unit outcrop positions; (2) provide a more complete section of the stratigraphic unit; (3) better preservation of contacts between units having significantly different resistances to weathering; (4) better preservation of delicate primary and soft sediment deformation structures in shale and siltstone units; (5) better preservation of trace fossils; (6) ability to obtain material for petrographic study below the present groundwater table; and (7) allows comparison of lithologic properties with petrophysical properties including wireline log responses. The advantages, however, are offset by the lack of a three-dimensional view, and the inability to observe directly lateral facies changes and large-scale sedimentary features. Specifically, planar and trough cross-stratification are not easily distinguished in core, but careful examination of details make distinction possible. Weimer and Tillman (1980) suggest that forests in troughs generally thicken in one direction or another; that they are repeatedly truncated at high angles through several feet of core; and that trough cross-strata commonly contain more random variations in grain size over short vertical distances.

One format that can be used for generating subsurface lithologic logs of continuous cores is shown in Figure 7. This graphical format follows that used by Casey (1980). Grain size, sedimentary structures, organic structures, and fossils are shown in the left hand column. Note that grain size increases to the left as does grain-size inferences made from SP and Gamma Ray wireline log curves. Lithologies and nature of the contacts between lithologies are presented in the central column.

<u>Wireline Logs</u>

Wireline logs were developed primarily for <u>in situ</u> assaying of a particular target zone (Merkel, 1979). In petroleum applications, this implies determining the amount of hydrocarbons in a stratigraphic unit. In uranium mineralization the average grade of uranium mineralization is determined. Generally, no single tool can give the desired results, therefore, more than one log is usually run in each borehole.

Wireline logs are also used to determine lithologies, and to interpretate depositional environments, correlate stratigraphic units, determine depths to a stratigraphic marker and stratigraphic, unit and measure sandstone thicknesses. In some applications, however, problems associated with the presence of authigenic (diagenetic) minerals exist.

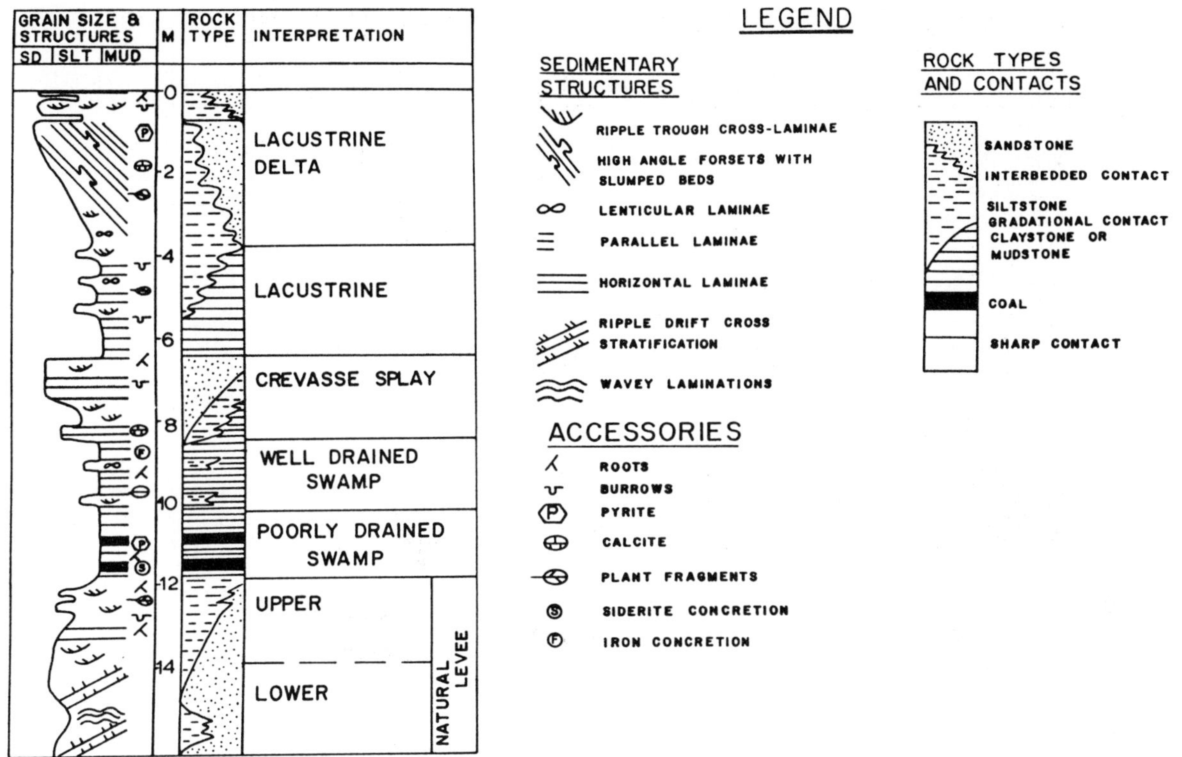

Figure 7. Idealized lithologic log and inferred environments of deposition, lower Wasatch Formation, Powder River Basin, Wyoming, illustrating a graphical method for displaying the vertical succession of grain sizes, sedimentary structures, lithologies and contacts in an outcrop or continuous core (after Ethridge, et al., 1981; modified from a method developed by A. J. Scott).

Idealized responses of the more commonly used wireline logs to various lithologies are shown in Figure 8.

Numerous authors have attempted to use log shapes or patterns from SP or Gamma Ray curves to infer vertical grain-size changes and hence depositional environments in sand-shale sequences (Krueger, 1968; Jagelaer and Matuszak, 1972; Selley, 1978 and Cant, 1984). The SP curve essentially records permeability, and because permeability is generally

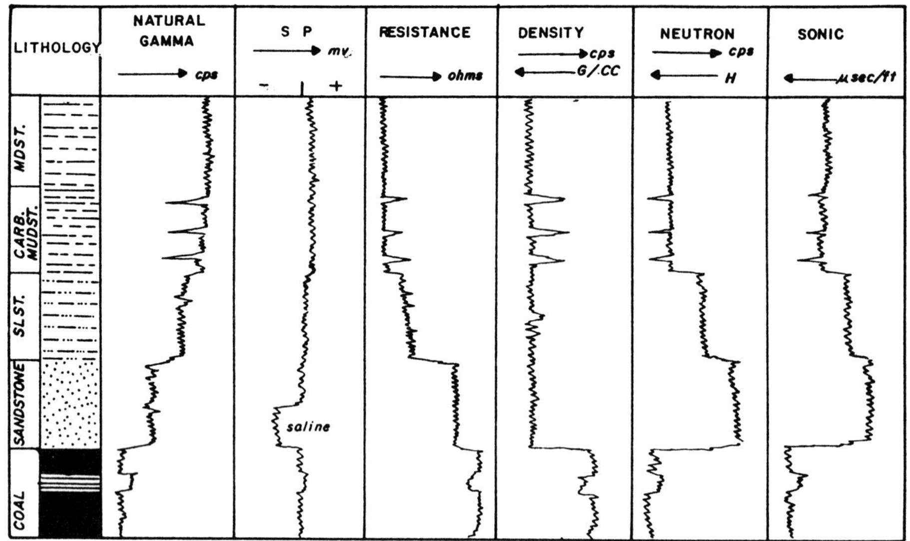

Figure 8. Wireline log responses to various lithologies associated with fluvial deposits (modified from Vaninetti and Thompson, 1982).

related to particle size in sandstone-shale sequences, the curve can be used as a continuous grain-size profile. The Gamma Ray curve records radioactivity of a stratigraphic interval and because this is essentially related to the amount of clay present, like the SP the Gamma Ray curve can be used as a grain-size profile (Selley, 1978). Problems in using the curves as continuous grain-size profiles stem from the presence of diagenetic clays and other minerals, abnormally high radiactive minerals such as mica, glauconite, zircon, etc., or hydrocarbons in sandstones. Several descriptive classifications for wireline log shapes hae been developed, and most are at least, vaguely similar to the ones presented in Figure 9. Remember that no log pattern alone can diagnose any particular depositional environment because of the problems listed above and because several different environments may be characterized by the same general vertical grain-size profile.

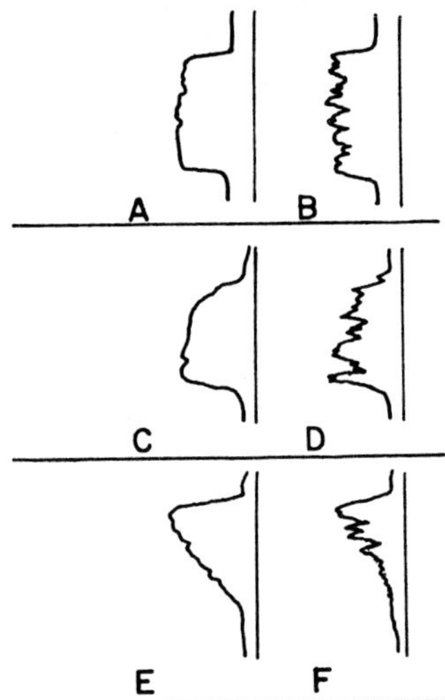

Figure 9. Typical end members of SP or GR log response in sand-shale sequence with emphasis on sand body: (A) cylindrical curve (inference; no preferred vertical grain size trend sharp bed boundaries); (B) as in A but with interbedded shales or tight streaks; (C) bell-shapped curve (inference; fining-upward grain size trend, gradational upper bed boundary and sharp lower bed boundary); (D) as in C with interbeds of shale or tight streaks; (E) funnel-shaped curve (inference; coarsening-upward grain-size trend, sharp upper bed boundary and gradational lower bed boundary); (F) as in E with interbeds of shale or tight streaks.

However, when coupled with core data, log patterns may provide useful data for interpretating depositional environments.

Rock-Log Calibration

It is essential to correlate continuous cores, closely spaced sidewall plugs, or cuttings with wireline logs before using the logs to infer depositional environments. First make certain that core depths tie to log depths. Commonly core and log depths differ, and a correction must be made to accurately correlate the two. Next compare the vertical sequence of grain size and sorting observed in core with the log shape. If the log shape does not conform to the observed size

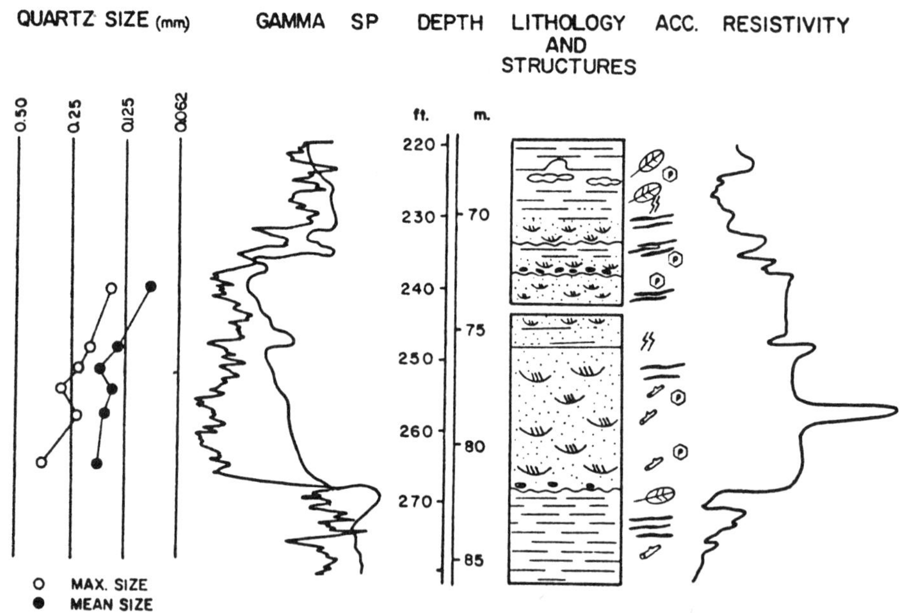

Figure 10. Detailed core description, wireline log responses and vertical quartz grain-size trends, Tertiary Hanna Formation, Hanna Basin, Wyoming. Illustrates core-log correlation techniques and problems (after Craig, et al., 1982).

trend (Fig. 10), look for tightly cemented zones, clay clasts, number and thickness of shale interbeds, and authigenic minerals. Finally, compare the types of contacts between lithologies with the log response. Once the core log correlations have been established within a local area, wireline logs from uncored wells may be evaluated with greater confidence.

Classification

Generally the most important feature of a fluvial system is the channel. The majority of channel classifications are based on the plan view morphology (Fig. 11). Single channel systems form a continuum from straight to highly sinuous (meandering). Braided systems exemplify a low sinuosity channel in which flow weaves around multiple, mobile channel bars at low-water stage. An anastomosing system is characterized by contemporaneous high-to-low sinuosity channels that weave around permanent, commonly vegetated islands or floodplain segments. Factors that favor one channel type over another, hydraulic and sedimentologic processes, and depositional models for each type will be reviewed in subsequent chapters. Schumm (1981; Fig. 12) presented a different classification of fluvial channels based on sediment load. His classification which divides channels into suspended load, mixed load and bedload types is used by Galloway in subsequent chapters of these notes. The relationship between channel type (based on sediment load) and channel pattern (plan view morphology) is also shown in Figure 12.

Acknowledgements

Chapters 2, 5 and 9 have been clarified by the comments and suggestions made by Steve Crews and Sally Carr. Many figures in these chapters have been drafted or redrafted by Herb Saperstone and Donna Upson. The chapters were typed by Jan Quintana and Beverly Everett.

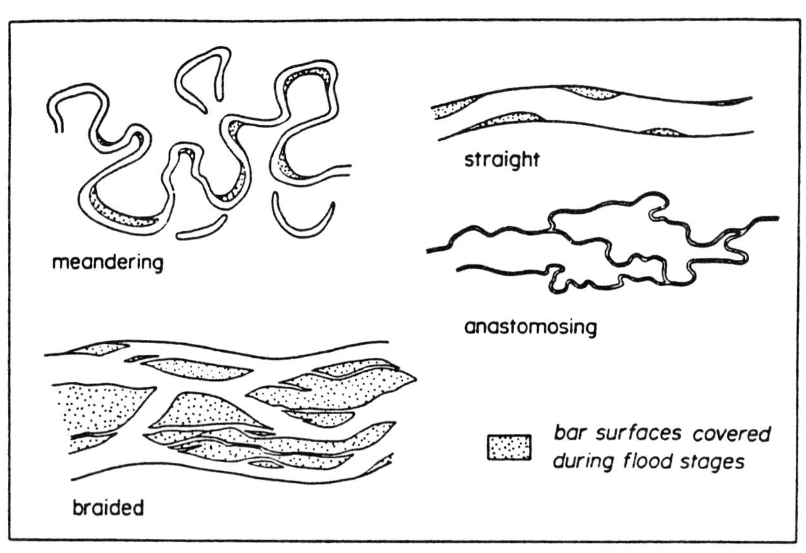

Figure 11. Plan view configuration of principal river types (after Miall, 1977).

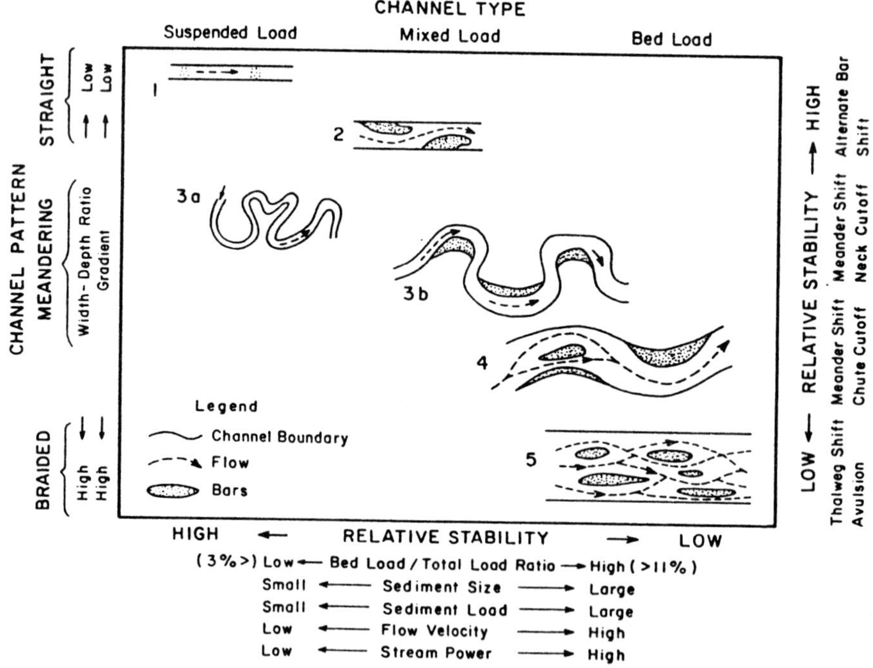

Figure 12. Channel classification based on pattern and type of sediment load with associated variables and relative stability indicated (after Schumm, 1981).

REFERENCES

Allen, J.R.L., 1966, On bed forms and paleocurrents: Sedimentology, v. 6, p. 153-190.

Allen, J.R.L., 1968, Current ripples: their relations to patterns of water and sediment motion: Amsterdam, North Holland Pub., 433 p.

Allen, J.R.L., 1970, Physical processes of sedimentation: Amer. Elsevier Pub. Co. Inc., New York, 248 p.

Brown, L.F., Jr., 1969, Geometry and distribution of fluvial and deltaic sandstone (Pennsylvania and Permian), northcentral Texas: Gulf Coast Assoc. Geol. Socs. Trans. v 19, p. 23-47.

Cant, D.J., 1984, Subsurface facies analysis, in, Walker, R.G. (ed.), Facies models (2nd Ed.): Geoscience Canada, Reprint Series 1, p. 297-310.

Casey, J.M., 1980, Depostional systems and basin evolution of the Late Paleozoic Taos Trough, northern New Mexico: Unpub. Ph.D. Dissertation, The Univ. of Texas, 236 p.

Collinson, J.D. and Thompson, D.B., 1982, Sedimentary structures: George Allen & Unwin, Boston, 194 p.

Craig, G.N. II, Burns, L.K., Ethridge, F.G., Laughter, T. and Youngberg, A.D., 1982, Overburden characteristics and post burn study at the Hanna, Wyoming underground coal gasification site: Stratigraphy, depositional environments and mineralogy: Hana Formation: U.S. Dept. of Energy, DOE/LC/10496-T1 (DE82009053), 188 p.

Davies, D.K., 1976, Models and concepts for exploration in barrier islands, in, Sacena, R.S. (ed.), Sedimentary environments and hydrocarbons: Amer. Assoc. Petrol. Geol. Short Course, p. 79-11

Davies, D.K. and Ethridge, F.G., 1975, Sandstone composition and depositional environments: Amer. Assoc. Petrol. Geol. Bull., v. 59, p. 239-264.

Davis, R.A., Jr., 1983, Depositional systems: A genetic approach to sedimentary geology: Prentice-Hall, Inc., New Jersey, 669 p.

Ethridge, F.G., Jackson, T.J. and Youngberg, A.D., 1981, Floodbasin sequences of a fine-grained meander belt subsystem: the coal-bearing lower Wasatch and upper Fort Union Formations southern Powder Basin, Wyoming, in Ethridge, F.G. and Flores, R.M. (eds.), Recent and ancient nonmarine depositional environments: models for exploration: SEPM Sp. Pub. 31, p. 191-209.

Ethridge, F.G. and Wescott, W.A., 1984, Tectonic setting, recognition and hydrocarbon reservoir potential of fan-delta deposits, in, Koster, E.H. and Steel, R.J. (eds.), Sedimentology of gravels and conglomerates: Can. Soc. Petrol. Geol. Memoir 10, p. 217-235.

Friend, P.F., Slater, M.J. and Williams, R.C., 1979, Vertical and lateral building of river sandstone bodies, Ebro basin, Spain: Jour. Geol. Soc. London, v. 136, p. 39-46.

Galloway, W.E. and Hobday, D.K., 1983, Terrigenous clastic depositional systems - applications to petroleum, coal and uranium exploration: Springer-Verlag, NY, 423 p.

Harms, J.C. Southard, J.B. and Walker, R.G., 1982, Structures and sequences in clastic rocks: Soc. of Econ. Paleo and Mineral. Short Course No. 9, (pages numbered by section).

Harris, D.G. and Hewitt, C.H., 1977, Synergism in reservoir management the geologic perspective: Jour. Petrol. Tech., v. 29, p. 761-770.

Jagelaer, A.H. and Matuszak, D.R., 1972, Use of well logs and dipmeter in stratigraphic trap exploration, in, King, R.E. (ed.), Stratigraphic oil and gas fields - classification, exploration histories: Amer. Assoc. Petrol. Geol. Memoir 16, p. 107-135.

Krueger, W.C., Jr., 1968, Depositional environments of sandstones as interpreted from electrial measurements - an introduction: Gulf Coast Assoc. Geol. Socs. Trans., v. 18, p. 226-241.

Merkel, R.H., 1979, Well log formation evaluation: Amer. Assoc. Petrol. Geol. Cont. Educ. Course Note Series No. 14, 82 p.

Miall, A.D., 1977, Fluvial sedimentology: Can. Soc. Petrol. Geol. Short Course Notes (pages numbered by section).

Miall, A.D. (ed.), 1978, Fluvial sedimentology: Can. Soc. Petrol. Geol. Memoir 5, 859 p.

Reading, H.D. (ed.), 1978, Sedimentary environments and facies: Elsevier, NY, 557 p.

Potter, P.E., 1967, Sand bodies and sedimentary environments: a review: Amer. assoc. Petrol. Geol. bull., v. 51, p. 337-365.

Reineck, H.E. and Singh, I.B., 1980, Depositional sedimentary environments: Springer-Verlag, NY, 549 p.

Rittenhouse, G., 1961, Problems and principles of sandstone-body classification, in, Peterson, J.A. and Osmond, J.C. (eds.), Geometry of sandstone bodies: Amer. Assoc. Petrol. Geol., p. 3-12.

Scholle, P.A. and Darwin, S. (eds.), 1982, Sandstone depositional environments: Amer. Assoc. Petrol. Geol. Memoir 31, 410 p.

Schumm, S.A., 1981, Evolution and response of the fluvial system, sedimentologic implications, in, Ethridge, F.G. and Flores, R.M. (eds.), Recent and ancient nonmarine depositional environments: Models for exploration, p. 19-29.

Selley, R.C., 1976, an introduction to Sedimentology (2nd. Ed.): Academic Press, NY, 417 p.

Selley, R.C., 1978, Concepts and methods of subsurface facies analysis: AAPG Cont. Educ. Course Note Series #9, 82 p.

Sherman, T., 1983, Sedimentology of the Upper Cretaceous Ericson Formation, Rock Springs Uplift, Wyoming: Unpub. M.S. Thesis, Colorado State Univ., 192 p.

Vaninetti, J. and Thompson, R.M., 1982, Geophysical well logging and related subsurface data useful in coal exploration and development: AAPG Short Course, Coal-bearing sequences - modern geological concepts for exploration and development, 34 p. + 85 figs.

Visher, G.S., 1965, Use of the vertical profile in environmental reconstruction: Amer. assoc. Petrol. Geol. Bull., v. 49, p. 41-61.

Walker, R.G. (ed.), 1984, Facies models (2nd Ed.) Geoscience Canada Reprint Series 1, 317 p.

Walther, J., 1884, Lithogenesis der Gagenwart. Beobachtungen uber die Bildung der Gasteine ander heutigen Erodberflache: Dritte Teil einer Einleitung in die geologie als historishe Wissenshaft, Jea: Verlag Gustav Fischer, p. 535-1055.

Weimer, R.J. and Tillman, R.W., 1980, Tectonic influence on deltaic shoreline facies, Fox Hills Sandstone, west-central Denver Basin: Colorado School of Mines Prof. Contr. No. 10, 131 p.

CHAPTER 3

ARCHITECTURAL-ELEMENT ANALYSIS: A NEW METHOD OF FACIES ANALYSIS APPLIED TO FLUVIAL DEPOSITS

Andrew D. Miall
Department of Geology
University of Toronto
Toronto, Ontario M5S 1A1 Canada

INTRODUCTION

The concept of the facies model has been the most powerful and successful tool devised by sedimentologists for classifying and explaining ancient sediments. At present there are at least a dozen formal fluvial facies models (Miall, 1980, 1981a), and many variants of these have been erected to explain specific ancient units. It has become clear that these models reflect fixed points on a continuum of variability. As discussed below, the continuum is, in fact, a multidimensional one because of the complexity of partly interdependent controls that govern fluvial sedimentation. A continuation of modelling studies along existing lines will simply result in a proliferation of arbitrary fixed points and a new approach is needed.

Friend (1983) proposed a classification of fluvial architecture based on a two-fold breakdown of the sediments into channel and interchannel sediments. Channels were further subdivided into fixed, mobile or sheet (i.e., non-channelized) types. Allen (1983) recognized "eight kinds of depositional features" or "internal architectural elements" in a study of a Devonian sandy braided stream deposit of the Welsh borders area. Ramos and Sopena (1983) defined five types of gravel and sand body in a Permo-Triassic unit in Spain.

These three studies contain the basis of a new architectural approach which, it is proposed here, can be applied to all fluvial deposits.

PROBLEMS WITH EXISTING METHODOLOGY

Facies models typically are constructed in the form of paleogeographic sketch maps, vertical profiles, block diagrams, or a combination of all three. These attempt to combine information on at least two scales: the assemblage of individual lithofacies units, and the geometry of mesoscale geomorphic elements such as channels and bars. Two interpretive features usually are emphasized: the characteristic vertical profile or cyclic sequence and, in the case of fluvial deposits, the morphology of the channels.

It has become clear that vertical profiles are not the rigorously diagnostic interpretive tools that they were once thought to be. Similar cyclic sequences can in many cases be produced in more than one way under the control of different autocyclic or allocyclic processes, and in different morphologic settings. This problem was discussed at some length by Miall (1980) and the ideas will not be repeated here. In addition, it has now been shown by careful study of large modern bedforms (Crowley,

1983), and well-exposed ancient sequences (Haszeldine, 1983a, b; Allen, 1983; Kirk, 1984) that interpretations based on vertical profiles can seriously misrepresent the geometry and complex internal structure of large macroform bar deposits.

Channel morphology has also been used as a primary key for interpreting fluvial sediments. The confusions between the ever-popular terms "braided" and "meandering" have been pointed out many times (e.g., Rust, 1978a; Miall, 1980) and Rust (1978a) proposed a more rigorous classification into four basic types: braided, meandering, anastomsing and straight, using quantitative sinuosity and braiding parameters. These four fixed points are useful simplifications, but several workers have illustrated spectra of channel morphologies that reveal a complete gradation between all four end members. For example, anastomosed rivers vary from highly sinuous to nearly straight (Smith, 1983). Lateral accretion, a process once thought to be characteristic of high-sinuosity "meandering" rivers, in fact also occurs in braided rivers (e.g., Bluck, 1979; Ori, 1979, 1982; Allen, 1983) and in some anastomosed reaches (Smith, 1983). High-sinuosity meandering rivers commonly contain numerous bars and islands, and thus show many of the characteristics of braided rivers (Miall, 1977; Jackson, 1978; Schwartz, 1978; Rust, 1978a; Forbes 1983).

The diversity of channel styles and deposit types exists because of the variety of partly interdependent controls that govern fluvial sedimentation. It is possible, conceptually, to isolate each one of these controls and vary its effects while keeping other controls fixed. A detailed discussion of this has been given elsewhere (Miall, 1980, Table 1).

BEGINNINGS OF A NEW APPROACH

Although facies studies are continuing to generate a wealth of variations on existing facies models, it is becoming clear that there are many points of similarity between fluvial deposits of all kinds. These can conveniently be considered under the heading of Jackson's (1975) classification of bedforms into microforms, mesoforms and macroforms.

Microforms are structures generated by turbulent variations in the inner part of the turbulent boundary layer. Small scale ripple marks and current lineations are the result. Such bedforms are essentially identical in all clastic environments and are thus non-diagnostic of fluvial style.

Mesoforms include the larger scale flow-regime bedforms, such as dunes and sandwaves, minor channels, and what Smith (1974) termed "unit bars", such as linguoid, transverse, longitudinal and diagonal bars. They are generated mainly by "dynamic events", particularly flood events occurring during storm-induced run-off or seasonal snow thaw. These bedforms and smaller bar forms also have similar geometries in all clastic environments under conditions of unidirectional aqueous flow.

Flow-regime bedforms (including the microforms) have essentially constant facies characteristics, and can therefore readily be described using a lithofacies classification scheme. That of Miall (1977, 1978; see Table 1) has been used, with minor modifications, for a wide variety of fluvial sediments. These facies characteristics can be identified in exposures in the order of a few metres across.

Determining the geometry of bar forms requires larger exposure, preferably three dimensional, and attention must also be paid to internal

Table 1: Lithofacies classification, from Miall (1978)

Facies Code	Lithofacies	Sedimentary structures	Interpretation
Gms	massive, matrix supported gravel	none	debris flow deposits
Gm	massive or crudely bedded gravel	horizontal bedding, imbrication	longitudinal bars, lag deposits, sieve deposits
Gt	gravel, stratified	trough crossbeds	minor channel fills
Gp	gravel, stratified	planar crossbeds	linguoid bars or deltaic growths from older bar remnants
St	sand, medium to v. coarse, may be pebbly	solitary (theta) or grouped (pi) trough crossbeds	dunes (lower flow regime)
Sp	sand, medium to v. coarse, may be pebbly	solitary (alpha) or grouped (omikron) planar crossbeds	linguoid, transverse bars, sand waves (lower flow regime)
Sr	sand, very fine to coarse	ripple marks of all types	ripples (lower flow regime)
Sh	sand, very fine to very coarse, may be pebbly	horizontal lamination, parting or streaming lineation	planar bed flow (l. and u. flow regime)
Sl	sand, fine	low angle ($<10^{\circ}$) crossbeds	scour fills, crevasse splays, antidunes
Se	erosional scours with intraclasts	crude crossbedding	scour fills
Ss	sand, fine to coarse, may be pebbly	broad, shallow scours including eta cross-stratification	scour fills
Sse, She, Spe	sand	analogous to *Ss, Sh, Sp*	eolian deposits
Fl	sand, silt, mud	fine lamination, very small ripples	overbank or waning flood deposits
Fsc	silt, mud	laminated to massive	backswamp deposits
Fcf	mud	massive, with freshwater molluscs	backswamp pond deposits
Fm	mud, silt	massive, desiccation cracks	overbank or drape deposits
Fr	silt, mud	rootlets	seatearth
C	coal, carbonaceous mud	plants, mud films	swamp deposits
P	carbonate	pedogenic features	soil

flow patterns, as deduced from paleocurrent studies. Therefore, satisfactory differentiation of even the smaller bar forms cannot be achieved from vertical profile studies of single outcrops or cores.

Macroforms reflect the cumulative effect of many dynamic events over periods of tens to thousands of years. They include major channels and the larger, compound bar forms such as point bars, side bars, sand flats and islands. It is the plan view of these macroform elements that generates the familiar fluvial channel styles, so commonly illustrated by low-level aerial photographs of modern rivers.

Superfically, these macroform elements would seem to define a very wide range of fluvial depositional styles, as referred to earlier in this paper. However, a major thesis of this paper is that at the scale of the smaller macroform elements (up to a few hundreds of metres in width and length) there are only about eight basic architectural elements, to use Allen's (1983) term. These elements are defined by grain size, bedform composition, internal sequence and, most critically, by external geometry (Fig. 1). The details of these characteristics vary, but it is suggested that all fluvial deposits are composed of varying proportions of these eight elements.

THE EIGHT BASIC ARCHITECTURAL ELEMENTS

Satisfactory definition of these architectural elements requires outcrops at least several tens of metres in width, mainly in order to reveal their cross-section geometry. The largest elements and those of sheet-like geometry may require hundreds of metres of lateral exposure.

Outcrops one or two orders of magnitude smaller than the scale of the element commonly cannot be properly identified because observations are limited to lithofacies assemblage and vertical profile, which commonly are non-diagnostic characteristics.

Recognition of the elements in the subsurface may be possible where a tight well network exists, particularly if cores are available. However, satisfactory diagnosis of architectural elements and river type cannot be achieved from isolated cores of outcrops.

Descriptions and definitions of architectural elements should include the following:

1. Nature of lower and upper bounding surfaces: erosional or gradational; planar, irregular, curved (concave or convex).
2. External geometry: sheet, lens, wedge, scoop, U-shaped fill.
3. Scale: thickness, lateral extent parallel and perpendicular to flow direction
4. Internal geometry: lithofacies assemblage, vertical sequence, presence of secondary erosion surfaces and their orientation, bedform paleoflow directions, relationship of internal bedding to bounding surfaces (parallel, onlap, downlap).

Many of these features are illustrated in Figure 1. Note that many lithofacies types appear in more than one element.

Markov chain analysis, a technique long advocated by the writer (Miall, 1973, 1977) may be useful for defining sequences within architectural elements, but it now seems much less useful as a general

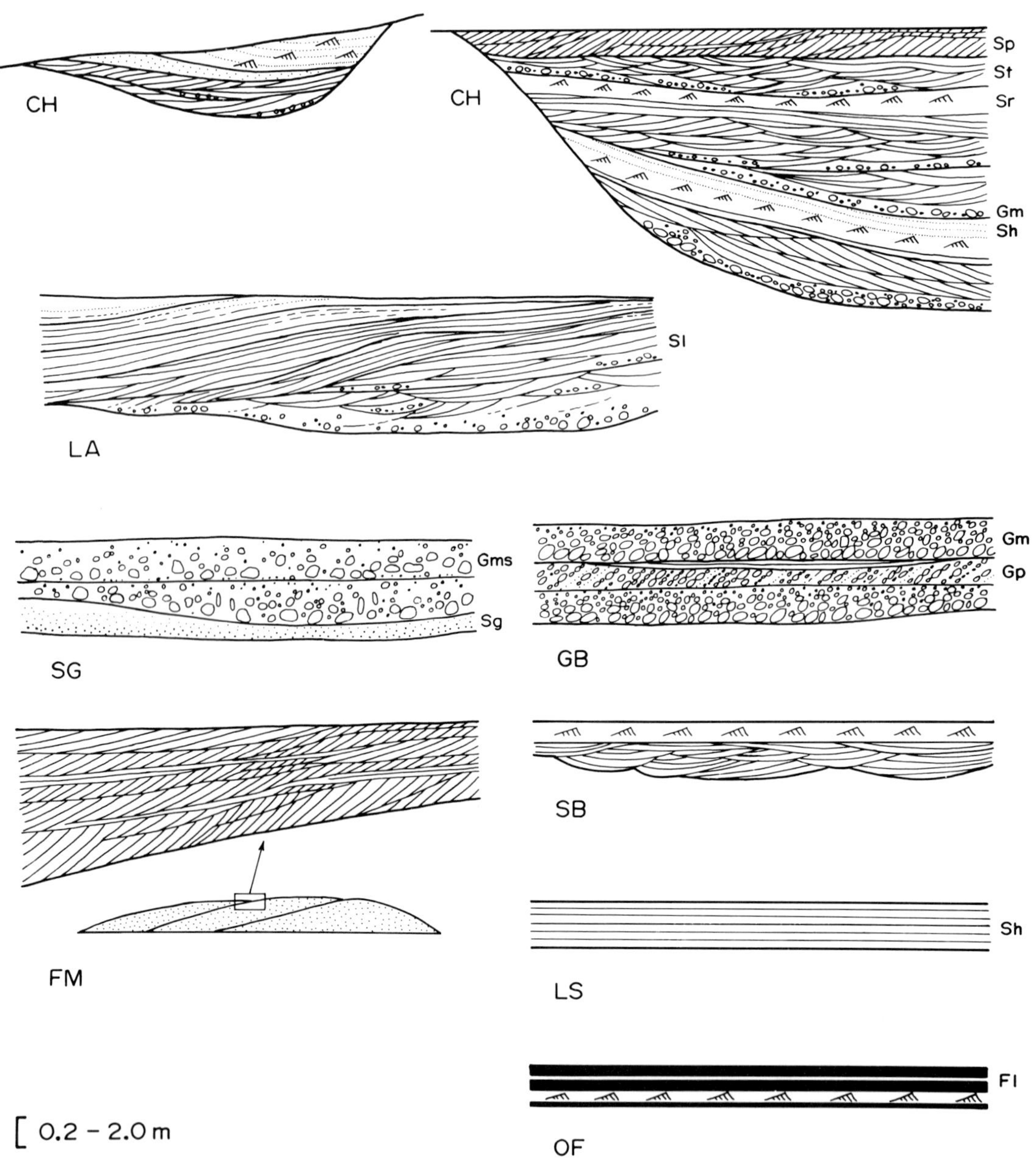

Fig. 1. The major architectural elements.
Vertical and horizontal scale are the same, and can vary, as shown.

analytical tool, because it cannot readily distinguish within-element from between-element facies superposition. Even where attention is paid to erosion surfaces (Cant and Walker, 1976; Miall and Gibling, 1978) the results may be of limited usefulness because of the variable significance of such surfaces.

The element hierarchy

The eight elements vary in scale and complexity. Smaller elements form stacked "storeys" (Friend et al., 1979) or "complexes" (Allen, 1983) within larger elements. The elements therefore form a hierarchy of scales, bounded by bedding contacts of variable significance. Allen (1983) defined a hierarchy of three types of bedding contact: first order contacts bound individual crossbed sets; second order contacts bound cosets (McKee and Weir, 1953) or genetically related lithofacies assemblages, such as some of the smaller elements of this paper; third order contacts define groups of elements or complexes, and usually are well-defined erosion surfaces. The basal scour surface of a major channel would constitute a third-order surface. Groups of channels, as in a paleovalley, would define an additional, fourth order. These concepts are illustrated in Figure 2.

Element CH: Channels

Channel geometry is conveniently defined by depth, width/depth ratio and sinuosity. The latter can rarely be observed in ancient rock units, except in cases of exceptionally good bedding plane exposure, but is

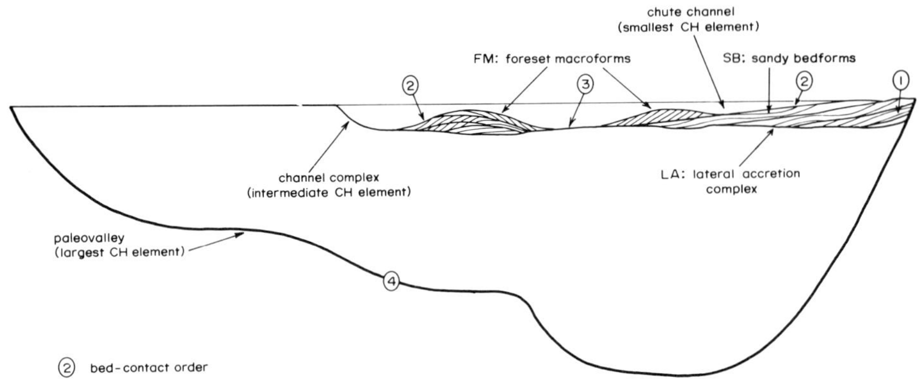

Fig. 2. The hierarchy of elements. Note nesting of channels, and nesting of bar complexes within channels. Recognition of this hierarchy depends on outcrop quality.

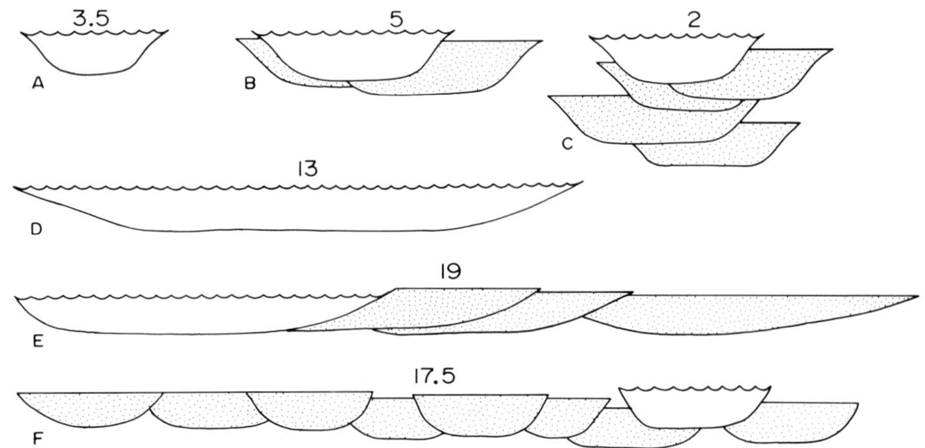

Fig. 3. Diagram to show the lack of relationship between the geometry of an individual active channel and the geometry of the resulting channel fill complex. Numbers above each channel complex are width/depth ratios calculated on the assumption that the active channel becomes completely filled with sediment and then switches to a completely different position. A,D: simple channels; B, E, F: broad channel-fill complexes formed by lateral channel migration or switching with little contemporaneous subsidence; c: stacked channel complex formed by vertical aggradation within a relatively stable channel under conditions of rapid subsidence.

commonly deduced on the basis of orthodox facies models assumptions (e.g., the assumed relationship between lateral accretion deposits and meandering rivers). This method of reasoning is to be avoided.

Major channels are rarely adequately exposed as they have widths of 10^3 to 10^5 m (Campbell, 1976, described an exceptionally well-exposed example). Their geometry may be reconstructed, given adequate lithofacies information, and this is one of the main objectives of fluvial facies analysis. Most of the remaining seven architectural elements are contained partly or entirely within the major channels, and these channels also contain a hierarchy of smaller channels (Williams and Rust, 1969; Rust, 1978a) which, because of their smaller size, are more amenable to field analysis.

The remainder of this section focusses on channels in the order of 10^1 to 10^2 m in width. These include the major channels of small to medium rivers and the minor channels of large rivers. The latter are second and lower order channels such as chute and bar-top channels and crevasse channels. They normally are initiated during high water stages, but may be incised and modified during falling water.

Channels may be classified into fixed (ribbon-shaped geometry) , mobile (broad and shallow with complex fill geometry) or sheet-like (essentially unchannelised), following Friend et al. (1979), Friend (1983) and Blakey and Gubitosa (1984). Fixed channels are narrow, with width/depth ratios less than 15. Mobile-channels are so called because they are filled by a process of channel migration or switching within a single major channel scour. Width/depth ratios are greater than 15. Where the width/depth ratio exceeds 100 the channel may be said to be sheet-like.

Channels have concave-up, erosional bases. The top of the channel fill may be erosional or gradational. Channels commonly have multistorey fills, with each storey bounded by an erosion surface. Channel margins become gentler in slope with increasing channel width. Slopes in excess of 45°, possibly even vertical or undercut, are not uncommon bordering narrow channels. Sheet-like channels may have practically imperceptible channel margins, sloping at a few degrees or less. These variations reflect bank stability. Channels cut into mud-dominated fines, particularly where the banks are stabilized by a dense root network, offer a considerable resistance to erosion (Smith, 1976) and tend to be steep. Those which cut into unconsolidated sand and gravel are easily eroded and may retreat rapidly, giving rise to lower channel margin slopes or stepped margins with steep cutbank sections alternating with flat terraces formed by bar complexes and partly filled minor channels. Where the sediment load of the river is dominated by sand or gravel a braidplain may develop, with almost unconfined, sheet-like channels. The channel cross-section geometry is therefore not necessarily an inherent property of a river with a particular slope discharge or sediment load, but at least partly reflects the nature of the pre-existing sediment into which the channel is cut (Crowley, 1983). This is one reason why attempts to define fluvial facies models on channel geometry have not always been successful.

Recognition of the channel-fill element in a fluvial deposit depends on the ability to define the sloping channel margins. This is commonly attempted by correlation of closely spaced outcrop or subsurface sections (e.g., Hopkins et al., 1982; Putnam, 1982a, b) but, because of the presence in most deposits of a hierarchy of channels of different scales, such correlation may be difficult or impossible (e.g., Wightman et al., 1981). Larger channel-fill complexes are better termed paleovalleys, and contain the accumulated deposits of many of the other types of element described later in the paper. They are bounded by fourth order bedding contacts (Fig. 2). Good examples up to 8 km across and 90 m deep are described by Blakey and Gubitosa (1984). Where the channel is of broad mobile or sheet type defining the channel margins may be difficult or impossible. Large channels filled by continually shifting minor channels (the familiar braided pattern) may contain evidence of several or numerous temporary channel margins, and the overall channel-fill geometry then means little in terms of conventional channel classifications (Schumm, 1963; see Fig. 3). Attempts to determine channel width and depth for the purpose of paleohydraulic reconstruction are likely to result in large errors.

If the channel margins cannot be defined field analysis is likely to result in a classification of the fill in terms of one or more of the other architectural elements. For example, the fill of ephemeral channels on arid braidplains, particularly on lake margins, are typically sheet-like and may consist mainly of elements SB: sandy bedforms, and LS: laminated sand sheets. Channels on the middle and upper levels of an alluvial fan commonly are filled by elements GB: gravel bedforms and SG: sediment gravity flows.

Channels filled by simple vertical aggradation commonly show fining-upward successions, reflecting one of two processes, progressive abandonment as a result of upstream avulsion, or the plugging action of a few dynamic events (e.g., flash floods). Typical cycles include

$$GB \rightarrow FM \rightarrow SB \rightarrow OF$$

$$LS \rightarrow SB \rightarrow OF$$

The thickness of such cycles cannot exceed the depth of the channel, and is likely to be much less where dynamic events strip away earlier deposits before depositing their sediment load.

Channels, particularly in high sinuosity systems, may be abandoned by chute or neck cut-off, in which case they will be filled by OF deposits showing a channelized, concave base.

Minor chute and bar-top channels, bounded by second order bedding contacts, contain assemblages of Ss, Se and Sl, with gravel lags and thin units of flow-regime bedforms (element SB: lithofacies St, Sp, Sh, Sr), showing no particular cyclic order.

Element GB: Gravelly bars and bedforms

Lithofacies Gm, Gp, Gt and Gl define a range of mesoforms. The simplest are the thin "diffuse gravel sheets" of Hein and Walker (1977), which are a few clasts thick, have diffuse, lobate margins, and move only during peak flow (lithofacies Gm). During episodes of high water and sediment discharge these sheets grow upwards and downstream by the addition of clasts, to form longitudinal bars (Rust, 1972; Hein and Walker, 1977). These bars reach about 1 m in height, and may show either an increase or decrease in clast size upward depending on their mode of accretion. Clast accumulation in place tends to result in an upward fining as the bar builds to shallower water levels. However, bars tend to fine downstream, and they may also migrate downstream. In such cases the coarser bar top migrates over the finer bar base (Gustavson, 1978).

Bars building into deeper water or areas of flow expansion, or bars covered by gradually waning flood events may develop lee-side separation eddies. This is accompanied by and encourages the growth of foresets leading to the development of transverse bars (lithofacies Gp). Hein and Walker (1977) proposed this evolutionary mechanism to explain the intimate relationship between lithofacies Gm and Gp, and the relationship has subsequently been confirmed by other workers (e.g., Gustavson, 1978; Massari, 1983).

Bluck (1979, 1980) showed that in some cases bars are capped by coarse gravels, which may interfinger with gravel or pebbly sand foresets, resulting in small coarsening upward sequences. Forbes (1983) referred to this as surface armouring. Such an arrangement may develop in several ways, such as the sweeping of gravel sheets across bar tops at high stage, and the development of sandy scour-fills at the toe of the foresets during lower water stages (Massari, 1983). Crowley (1983) showed that similar upward-coarsening textures occur in some large sandy bar forms and are the product of changing water velocity and depth over the bar crest during active bar growth (see element FM, below). Coarsening-upward, therefore, is probably a dynamic component of many large bar elements.

Lithofacies Gt represents migration of transverse bars with curved crest lines, or the fill of minor channels. Where such channels debouch into pools they developed crossbedded chute bars (lithofacies Gp) (Ramos and Sopena, 1983; Massari, 1983). In rare cases lateral accretion sets can be recognized. Such deposits are defined as a separate element because of their implication for relatively long term lateral migration of channel-bar complexes, resulting in a distinctive architecture.

Element GB typically forms multistorey sheets tens to hundreds of metres thick. Flat or irregular erosion surfaces between bar sets are common. Steeply dipping channel margins are rarely seen, partly because they tend to be minor parts of a gravelly fluvial landscape. Actively migrating channels may undercut older bar gravels producing cutbanks 1 or 2 m high, but when filled with later bar gravels of similar composition and texture the cutbanks may be very difficult to identify.

Element GB may be interbedded with minor to predominant sheets or lenses of element SG: sediment gravity flows. Element SB typically comprises at least 5% to 10% of even the coarsest gravel succesion, and represents slack water deposits, such as abandoned-channel fills (minor element CH where identifiable) or bar-edge sand wedges and microdeltas (Rust, 1972; Miall, 1977). Downstream, element GB usually is progressively replaced by elements SB and FM (Miall, 1978; Vos and Tankard, 1981; Brady, 1984).

Element SB: Sandy bedforms

The familiar flow regime bedforms that form in sand-dominated river systems have been described by many writers (Southard, 1971; Allen, 1968; Miall, 1977; Harms et al., 1975, 1982). Dunes (lithofacies St), sand waves, linguoid and transverse bars (lithofacies Sp), upper flow regime plane beds (lithofacies Sh) and ripple marks (lithofacies Sr) occur in a wide variety of fluvial settings and show a range of assemblages and vertical sequences.

In some cases large exposures show that short sequences of bedforms are interbedded with each other over wide areas below major convex-up bedding contacts, indicating that they were dynamically related and formed simultaneously. This type of architecture is the key diagnostic characteristic of element FM: foreset macroforms, to be described below. A special type of macroform is the lateral accretion element (LA), distinguished by the presence of lateral accretion surfaces. This element is also described below. Both these complex elements are bounded by second order bedding contacts.

Where these architectural features can be conclusively ruled out the deposits probably represent fields or trains of individual bedforms. Vertical stacking of different bedform types indicates long or short term changes in flow regime. Short term changes occur during stage changes (flash floods, seasonal fluctuations). Longer term changes reflect aggradation and reduction in water depth. Both can result in similar lithofacies assemblages and sequences (which is one of the problems with vertical profile analysis) requiring examination of the architecture and overall context of the deposits in order to arrive at correct interpretations. Such deposits contain first and third order bedding contacts, but most lack the second order contacts that define macroform complexes.

Examples of the SB element include fields of dunes (lithofacies St) in the deeper portions of active channels and fields of transverse bars or sand waves (lithofacies Sp) on the tops and flanks of point bars and sand flats.

Miall (1977) defined a distinctive lithofacies assemblage dominated by Sp and showing little or no internal cyclicity. This was named the

Platte-type of braided river deposit, after the Platte river, Nebraska. The original interpretation of this assemblage was that it represented the migration of fields of linguoid or transverse bars, many of which were capped or draped by Sr or Fl during falling water (Smith, 1970, 1971, 1972; Blodgett and Stanley, 1980). The "crossbedded simple bars" of Allen (1983) are similar. However, Crowley (1983) has shown that in at least some cases the linguoid bars form part of much larger macroform structures 200-400 m in length and 0.7 to 1.5 m high. The architecture of these structures is discussed in the next section.

Many workers have described the characteristic small scale crossbedding that occurs in shallow areas of active channels, particularly on bar tops. Various types of ripple cross-lamination (lithofacies Sr) are the result. These small scale structures typically are deposited during falling water and, where preserved, their capping of larger bedforms or bars produces local fining-upward sequences. Such sequences are almost ubiquitous in fluvial environments and their occurrence has little diagnostic value.

Crevasse channel and crevasse splay deposits typically are composed of element SB.

On distal braidplains, such as those bordering playa lakes, fluvial deposits may be entirely composed of element SB. Sheets of sand develop in broad, virtually unconfined channels. Aggradation and progressive abandonment of these channels occurs slowly or during single flood events. In either case fining-upward cycles are commonly the typical result. Similar deposits characterize the arid "terminal fan" deposits of northern India (Parkash et al., 1983).

In the pre-Devonian, the lack of vegetation is thought to have resulted in a predominance of weakly-channelized bedload streams (Schumm, 1968). The architecture and composition of the resulting fluvial deposits probably was in many cases similar to the distal braidplain sand sheets described here. Long (1978) discussed some Proterozoic examples.

Particularly vigorous flood events in ephemeral channels may produce a distinctive type of lithofacies assemblage and sand body geometry, described below under the heading of element LS.

Element FM: Foreset macroforms

For no other type of fluvial deposit is three-dimensional architectural analysis more essential than in the case of the macroform elements described in this section. Large compound bar forms have been described from many modern rivers, including the side bars of the Tana (Collinson, 1970) and the sand flats of the South Saskatchewan (Cant and Walker, 1978). Many such studies have been reported by Bluck (1976, 1979). (Point bars and related bank-attached forms are also included in the macroform category, but have received much greater attention from sedimentologists and are discussed in a separate section.)

It is only recently that studies of a few selected ancient sequences have begun to examine the internal geometry of these major components of the fluvial environment. Vertical profile analysis is quite inadequate for this purpose, and so it is unlikely that useful contributions are to be made to macroform study using only subsurface core and log data, except in areas of unusually close well spacing. Macroforms may be up to a kilometre across and contain a complex internal geometry that can only be elucidated by the study of large open-cast mine faces, road cuts or natural cliffs. These deposits represent the most vigorous depositional activity of fluvial flow systems, and their analysis is therefore crucial to a correct interpretation of fluvial style.

The essential characteristics of a foreset macroform are that it consists of several (posibly numerous) cosets of flow regime bedforms dynamically related to each other by a hierarchy of internal bounding surfaces (Fig. 4). These reveal the former existence of an active, non-periodic, possibly irregularly shaped bar form comparable in height and width to the channel in which it formed. The few studies of ancient macroforms that are available reveal very few similarities in detail, suggesting that there is a fertile field of research here for investigating the relationships between flow width and depth, discharge amount and variability, sediment grain size, and the composition and geometry of the resulting deposit. One of the few points of similarity is the presence in at least some of the published examples of second order bounding surfaces dipping gently downstream (Haszeldine, 1983a, b; Kirk, 1983; Allen, 1983) or, gently upstream around and over a low relief bar core (sand shoals of Allen, 1983). Between these bounding surfaces are sets or cosets of St, Sh, Sl or Sr. The Sh and Sl laminae are organized parallel or subparallel to the second order bounding surfaces. Detailed paleocurrent studies show that the flow regime bedforms advanced generally down the slopes defined by the downcurrent-dipping second order surfaces (Haszeldine, 1983a, b; Kirk, 1983) or oblique to the surfaces draping bar cores (Allen, 1983). These data reveal a picture of fields of bedforms driving across, around and down the bar forms. Flow-transverse bedforms such as the cross-channel "bars"

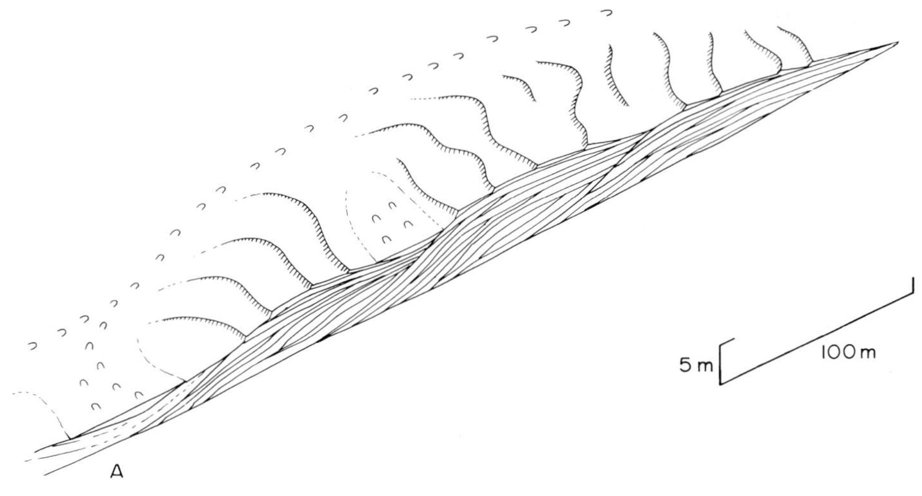

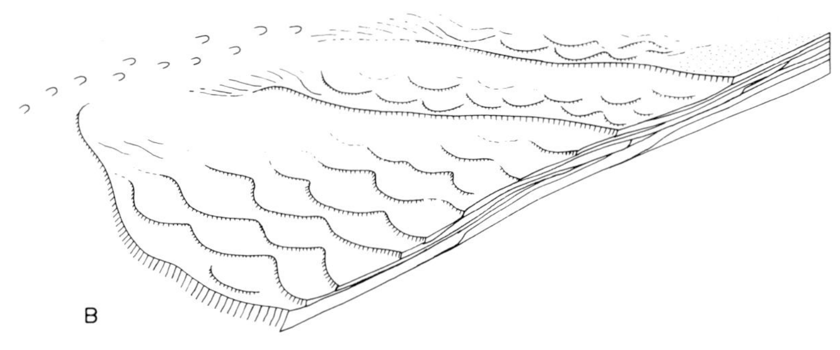

LEGEND
for Figs. 4, 7-18

- dunes (large-scale 3-D ripples)
- crestlines of straight to sinuous ripples
- gravel riffle
- exposed but intermittently active bar top
- swamp
- cutbank
- incipient vegetation
- mature vegetation

Fig. 4. Examples of foreset macroform elements. A). Loosly based on Allen (1983) and Kirk (1983), B). Loosly based on Cant and Walker (1978) and Haszeldine (1983a, b). Scales are approximate. Macroform geometry and internal structure vary considerably depending on channel depth, grain size, discharge amount and variability.

of Allen (1983) and Cant and Walker (1978) may move more slowly over the crest of the bar, and may become anchored completely if an emergent nucleus is present. The opposite end of the crestline, in deeper water, continues to advance more rapidly, so that the entire bedform swings around up on to the bar core (Cant and Walker, 1978; Allen, 1983, Fig. 19). The macroforms accrete sediment partly by this process of bedform capture on the upstream or flanks, and partly by rapid burial and preservation of superimposed bedforms on the advancing downstream face.

Many of the variations in composition and geometry between described macroforms probably reflect fluctuations in stage. Many of the first and second order bounding surfaces have the character of reactivation surfaces (Collinson, 1970). The "sand flat" macroforms of Cant and Walker (1978) are cut by numerous erosional channels during falling water. Kirk (1983) described a distinctive low-stage lithofacies assemblage draping the macroform, distinguished from the body of the structure by divergent paleocurrents that reflect falling-water surface run-off and bar-top channel orientation.

Descriptions of macroforms in modern rivers suffer from the lack of three-dimensional control. Thus Cant and Walker's (1978) sand flat model (their Fig. 14) predicts a simple tabular sheet of Sp cosets. Crowley (1983) described a Platte-type macroform consisting of a single large scale Sp set resting on an apron of fines and draped by coarser grained St or Sr sets. The upward coarsening reflects varying shear stress in relation to increasing water depth from top to bottom of the advancing foresets. Missing from these descriptions are any indications of second order internal bounding surfaces. This missing information is reminiscent of Jackson's (1978) inability to locate epsilon crossbedding on many modern point bars, and points to a common problem with facies studies based on the modern record.

Element LA: Lateral accretion deposits

Where the main flow in a channel is directed away from the bank, as on the inside of a curve, centrifugal forces lead to the development of a helical overturn and a secondary current passing obliquely up the bed of the inner bank. Because of the reduced shear stress associated with this current, significant sedimentation takes place, and the bank accretes laterally at a high angle to the principal flow direction. A distinctive architectural element results, characterized by large scale, gently dipping second order bounding surfaces that correspond to the successive increments of lateral growth. These dipping surfaces are traditionally termed epsilon crossbedding (after Allen, 1963, 1965). They usually show offlapped upper terminations, followed by fine-grained facies of the OF element. Their lower terminations downlap onto the channel floor. The base of a LA element is therefore erosional and the top gradational, except where truncated by a younger element. The height or thickness of the element approximates the bankfull depth of the channel. Recognition of the LA element can therefore be an important first step in paleohydraulic analysis.

The internal geometry and lithofacies composition of the LA element is highly variable, and depends on channel geometry and sediment load. The width is approximately two-thirds of the channel width (Allen, 1965), at least in single-channel rivers, so that the dip of the lateral accretion surface varies according to width/depth ratio (Leeder, 1973). With a width/depth ratio of 3, epsilon dip may reach 14°, whereas with a width/depth ratio of 80, the dip theoretically is as low as 1°. However, in wide channels a simple, gently dipping bank surface is unlikely; the inner bank of a bend is typically covered in bars and chute channels, obscuring the simple geometry of the LA element (e.g., Schwartz, 1978). Inside the bends of single-channel, high-sinuosity rivers laterally extensive LA deposits 10^2 to 10^3 m across, termed point bars, are typically developed. Within low sinuosity (e.g., braided) rivers the LA element is less prominent, but may even occur in straight channels where alternate bars develop inside the meanders of a sinuous thalweg (e.g., Smith, 1983).

Lithofacies assemblages within LA elements vary markedly, depending on the calibre of the sediment load, and on discharge variability. Gravel-dominated deposits are relatively rare, and in most gravelly fluvial deposits are subordinate to element GB. Deposits consisting of sand or pebbly sand contain a wide variety of lithofacies reflecting vigorous bedform and bar progradation and chute development. Bedding within this type of LA element is complex and may obscure the underlying laterally accreted geometry. Indicated flow directions in these crossbedded deposits are parallel or subparallel to the strike of the epsilon sets. The simplest LA elements are those composed of fine sand, silt and mud. Secondary bedforms on the accretionary surfaces are rare and small in scale, and the epsilon sets are relatively steeply dipping and readily identified in outcrop.

LA deposits do not retain a constant geometry or composition around any given meander bend. As a result, the classic fining upward profile (Allen, 1970) may not be present. In gravelly rivers Bluck (1971) and Bridge and Jarvis (1976) showed that the coarsest part of the point bar is located at the upstream end of the bar (bar head) and may migrate downstream over sandy bar tail deposits. Jackson (1976a) found that in the Wabash River (sand and pebbly sand) the helical flow patterns responsible for the fining-upward point bar profile tend to develop only in the downstream part of a meander bend. Nanson and Page (1983) showed that within tight meanders flow separation may occur at the downstream end of a point bar. Eddy currents there form significant deposits of fine sand, silt and mud in "concave bench complexes".

Lateral accretion deposits can be classified into four groups, according to grain-size: gravel, gravel-sand, sand, and sand-silt-mud. Gradations between these groups are to be expected, even within a single fluvial deposit, as a result of variations in the energy of discharge events, plus longer term changes in tectonic and climatic control.

Examples of these four types are shown in Figure 5.

Element SG: Sediment gravity flow deposits

This element occurs as narrow, elongate lobes or multistorey sheets, and is typically intimately interbedded with element GB (Fig. 6). The predominant lithofacies is Gms.

This element is formed primarily by debris flow and related sediment gravity flow mechanisms. Individual beds average 50 cm to 3 m in thickness, rarely exceeding 3 m. Flow units may be lobate in plan view, with widths of up to about 20 m and downstream lengths of several kilometres (data from Hooke, 1967; Wasson, 1977; Vessell and Davies, 1981; Nemec and Muszynski, 1982). Amalgamated flows with total thicknesses of several metres are common.

Flow units typically have irregular, non-erosive bases. Plow events

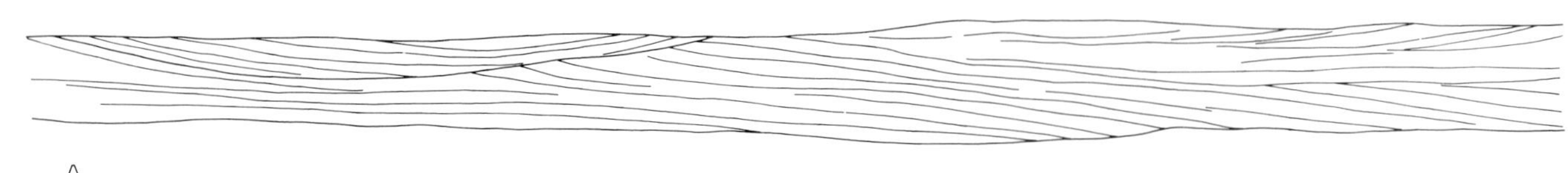

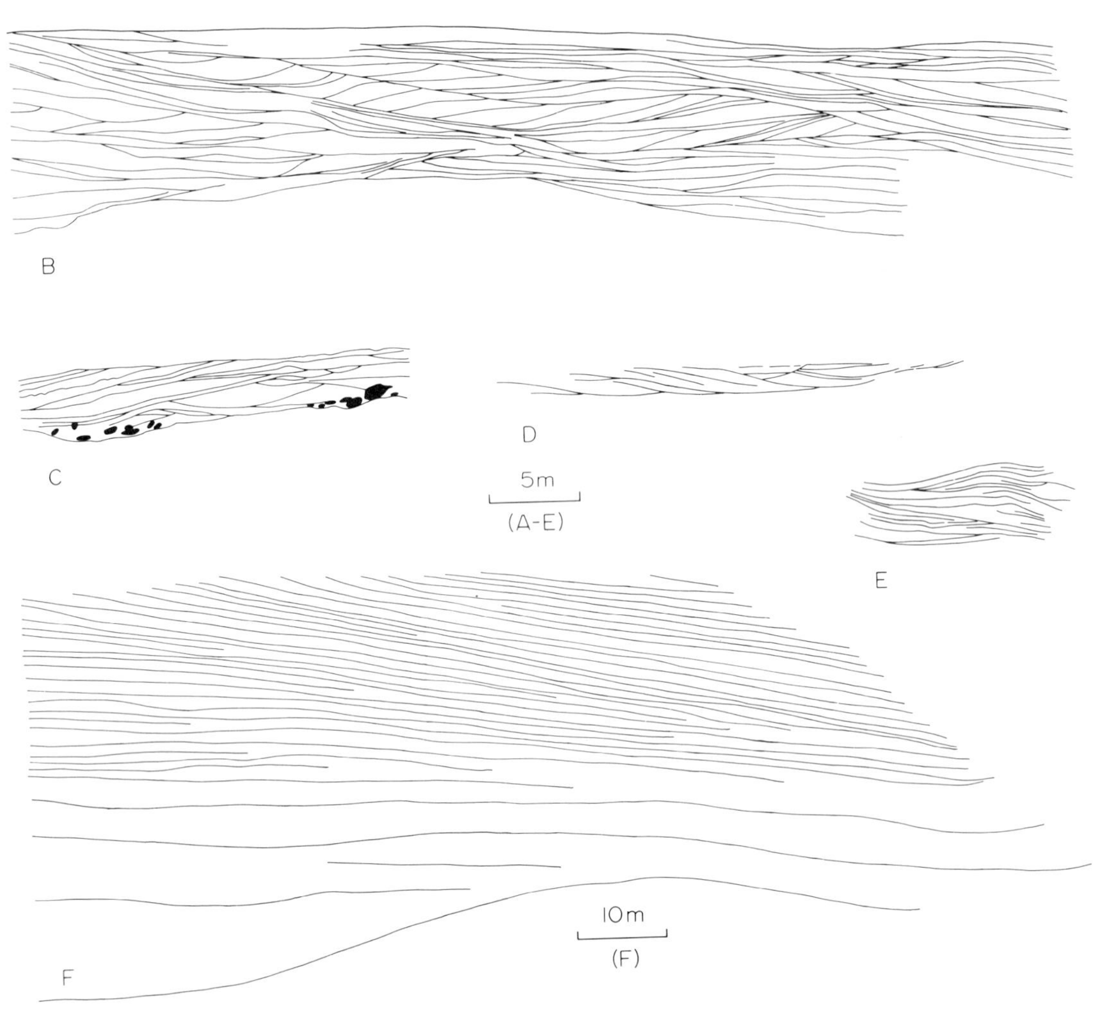

Fig. 5. Examples of lateral accretion elements. Fluvial model numbers are those in Table 2, and are discussed in text. A). Conglomeratic point bar (lithofacies Gm), with chute channel (lithofacies Gt). Fluvial model 4 (Ori, 1979); B). Element composed of medium grained sandstone, with abundant internal planar-tabular crossbedding (lithofacies Sp). Fluvial model 6 (Beutner et al., 1967); C). Fine to very coarse sandstone and pebbly sandstone, with cobble to boulder conglomerate lag. Abundant internal crossbed structures (lithofacies Sp, St, Sh and Sl. Fluvial model 5 (Allen, 1983); D. Small sandy point bar with abundant dune and ripple crossbedding (lithofacies St, Sr). Fluvial model 6 (Puigdefabregas, 1973); E). Point bar composed mainly of fine sandstone and siltstone (lithofacies Sl), with minor medium to coarse crossbedded sandstone (lithofacies St) at base. Fluvial model 7 (Nanson, 1980). F). Giant point bar with thick bedded, fine grained, trough crossbedded sandstone at base (lithofacies St), passing up into epsilon set of fine sandstone and argillaceous siltstone (lithofacies Se). Fluvial model 6 (Mossop and Flach, 1983).

passively occupy existing erosional channels or the irregular topography formed by earlier sediment gravity flow and sheet flood events. Internally they may show a wide range of textures and fabrics. Disorganized textures are typical of rigid plugs that are rafted at the centre of some debris flows (Bull, 1977). Grading and inverse grading are common. Nemec and Muszynski (1982) described an upward transition in some flow types (their Facies C) from graded to low-angle cross stratified gravels, which they interpreted as a transition from debris flow to traction transport mechanisms. Buck (1983) described a sand dominated diamictite facies interpreted as "mud flow" deposits. Shultz (1984) proposed a four-fold classification of debris flows based on matrix content, packing characteristics and grading. He showed that these can conveniently be described in the field using the diamict lithofacies code scheme of Eyles et al. (1983).

Element LS: Laminated sand sheets

Sheets of laminated sand (lithofacies Sh, Sl) with minor Sp, St or Sr are common to dominant in some ancient rock sequences, and have been interpreted as the product of flash floods depositing sand under upper flow regime plane bed conditions (Miall, 1977, 1984; Rust, 1978; Tunbridge, 1981; Sneh, 1983). The flood deposits of Bijou Creek, Colorado, are invariably quoted as a close modern analogue (McKee et al., 1967); they provided the basis for the Bijou Creek fluvial model of Miall (1977). Ephemeral streams of the Lake Eyre Basin also contain local accumulations of this assemblage (Williams, 1971).

The characteristic architecture of this element has been best described by Tunbridge (1981) and Sneh (1983). Individual sand sheets are 0.4 to 2.5 m thick, and rest on flat to slightly scoured erosion surfaces. They may be capped,gradationally, by Sp, St or Sr, indicating waning flow conditions at the end of a flood event. Individual sheets may be traced laterally for more than 100 m. At the edges they thin and split into thinner units dominated by finer grained sands and silts of lithofacies Sr. These beds probably represent the margins of individual flood sheets. Channel cutbanks are rare to absent. Stacked sequences may reach tens of metres in thickness.

Element OF: Overbank fines

Friend (1983) has pointed out that there are many factors which control the geometry and thickness of overbank sequences and their relative importance in a fluvial succession. Among these are sediment supply, channel pattern, subsidence rate and channel migration/avulsion behaviour. Therefore, as with all the fluvial deposits described here, correct interpretation can only proceed from carefully documented architectural descriptions.

Element OF is characterized by lithofacies Fl, consisting of mud or silt with thin lenses or laminae of silt to fine sand, commonly showing ripple cross-lamination. Additional facies may include floodplain pond muds with freshwater molluscs, coal, calcrete, and crevasse splay sand sheets. The latter are described briefly under the heading of element SB, above. Mapping of calcrete or tuff horizons in this element may provide useful marker horizons for linking isolated field sections, and thus

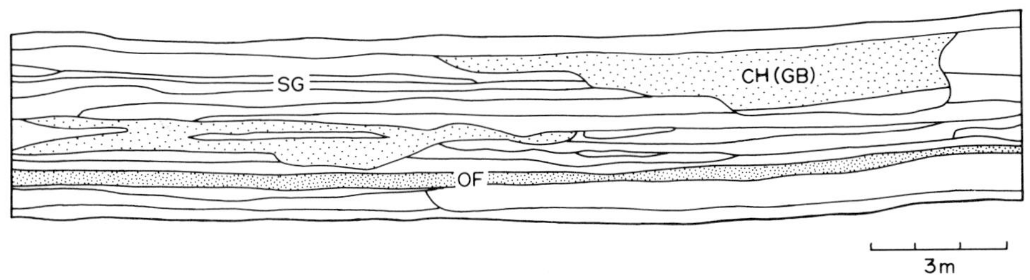

Fig. 6. Interbedding of elements GB and SG (Wasson, 1977).

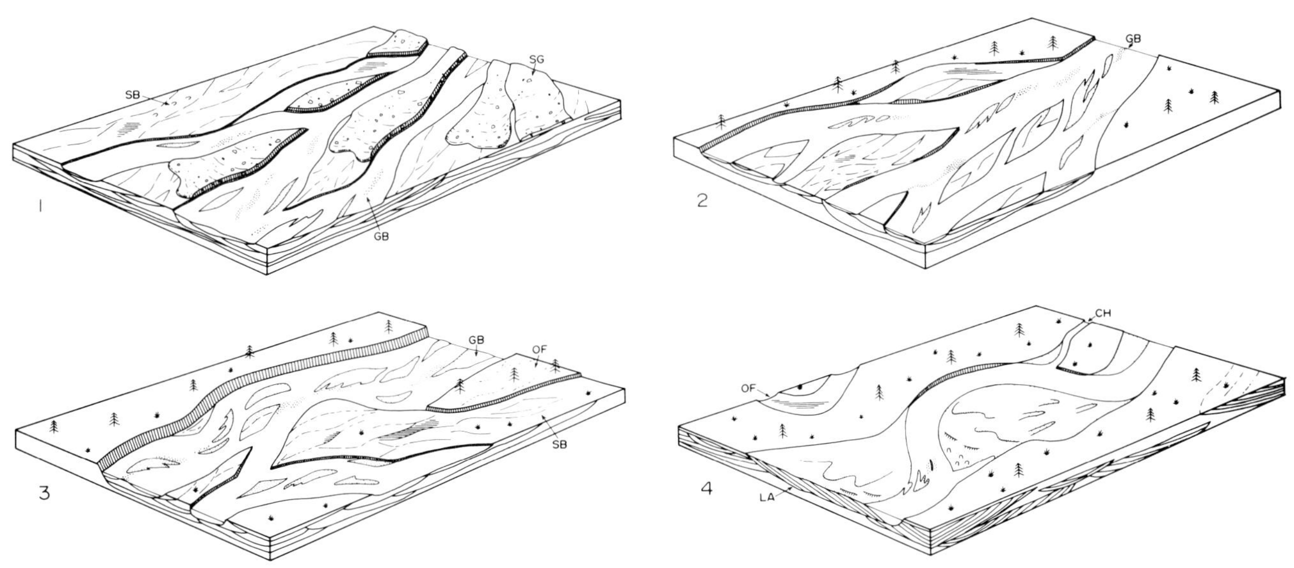

Fig. 7. Model 1: proximal alluvial fans with sediment gravity flow lobes.

Fig. 8. Model 2: proximal alluvial fan or outwash braidplain.

Fig. 9. Model 3: major gravelly, low sinuosity river with well-defined topographic levels.

providing much needed three-dimensional architectural control (Allen, 1974; Allen and Williams, 1981, 1982).

Most OF deposits have a sheet-like geometry, reflecting their origin by vertical aggradation. Near active channels the sheets are split by crevasse splays and display the low depositional dips of levees. They are truncated abruptly by channel cutbanks. OF deposits may fill abandoned channels, in which case they show the characteristic concave-up basal contact and ribbon to lensoid geometry of the channel itself (e.g., Ethridge et al., 1981, Fig. 11).

EXAMPLES OF ARCHITECTURAL STYLE

For rivers, as for buildings, architectural style has been categorized by the use of standard terms (Romanesque, Gothic, Baroque, etc.) but, as noted earlier, the fluvial equivalents (braided, meandering, Platte-type, etc.) are rapidly ceasing to be useful because of the variety of controls which can, incombination, lead to the development of any particular style.

The purpose of this section is to illustrate the geomorphology and fluvial architecture of a range of fluvial styles by means of a suite of block diagrams. Some are familiar and much quoted models, others are based on very few studies of the modern or ancient record, and are therefore less well known (Figs. 7 to 18, Table 2).

It must be emphasized that this section does NOT represent an attempt to provide a comprehensive suite of fluvial models, to replace or add to those already in existence. The purpose is solely to illustrate some of the variability in style that is possible in natural environments, much of it barely appreciated by fluvial sedimentologists. Many more local models will undoubtedly be erected once we free ourselves from the constraints of the existing suite of standard types. Every conceivable gradation between any two of the models illustrated here is to be expected, and the numbers used to identify the models are only for the convenience of the reader of this chapter.

Model 1 is characteristic of proximal regions of alluvial fans where the balance between source area weathering rates and rainfall results in abundant debris flows. Rivers draining active volcanic regions also may show this fluvial style. Lobate SG units are interbedded with channelized or tabular sheet-flood beds of GB. Minor units of SB may occur in abandoned channels (Figs. 6, 7). Hooke (1967) has shown that the frequencey of debris flows depends strongly on source rock weathering characteristics, so that adjacent fans, the headwaters of which flow across contrasting bedrock units, may show quite different lithofacies assemblages.

Model 2 occurs within alluvial fans where debris flows are sparse. The proximal reaches of outwash braidplains also form deposits of this type (e.g. Boothroyd and Nummedal, 1978). The valley is crossed by numerous broad, shallow channels of low sinuosity, that branch and join, and constantly change in position as a result of cutbank erosion and bar progradation (Fig. 8; Bluck, 1979). The bulk of the deposits (typically up to 95% of total thickness) are tabular bodies of element GB, including diffuse gravel sheets and longitudinal bars (Gm), transverse gravel bedforms (Gp) and scour fills (Ge, Gt). During stage fluctuations bar complexes become emergent, and are crossed by minor channels within which

Table 2: Architectural Characteristics of some common fluvial styles

No.	Sinuosity	Braiding Parameter	Sediment Type	Characteristic Elements	Examples
1	low	high	gravel, minor sand	GB, SG (SB)	Hooke (1967), Wasson (1977), Nemec and Muszynski (1982)
2	low	high	gravel, minor sand	GB (SB)	Boothroyd and Nummedal (1978), Vos (1981), Miall and Gibling (1978)
3	low to int.	int. to high	gravel, minor sand, fines	GB, SB (OF)	Williams and Rust (1969), Rust (1972, 1978b), Steel (1974), Minter (1978), Miall (1984b)
4	high	low to int.	gravel, minor sand, fines	GB, LA, OF (SB)	Bluck (1971), Gustavson (1978), Schwartz (1978), Ori (1979, 1982)
5	int.-high	low to int.	sand, pebbly, minor fines	SB, LA, OF (GB)	Allen (1983), Levey (1978), McGowen and Garner (1970), Nijman and Puigdefabregas (1978), Crowley (1983)
6	high	low	sand, minor fines	LA, SB, OF	Beutner et al. (1967), Nami and Leeder (1978), Puigdefabregas and Van Vliet (1978), Plint (1983), Stear (1983)
7	high	low	fine sand, silt, mud	LA, SB, OF	Miall (1979), Nanson (1980), Jackson (1981), Stewart (1983)
8	low to high	high	sand, fines	SB, OF (LA)	Smith (1983), Rust (1981)
9	low to int.	high	sand	SB, FM	Miall (1976, 1984b), Blodgett and Stanley (1980), Crowley (1983), Allen (1983)
10	low to int.	int. to high	sand, minor fines	FM, SB, OF	Cant and Walker (1978), Kirk (1983), Haszeldine (1983a,b), Allen (1983)
11	low	high	sand, minor fines	SB (OF)	Williams (1971), Miall and Gibling (1978)
12	low	high	sand, minor fines	LS (OF)	McKee et al. (1967), Rust (1978b), Miall and Gibling (1978), Tunbridge (1981), Sneh (1983), Miall (1984b)

Element CH not differentiated, but present on a variety of scales in all examples.
Elements shown in brackets are minor components.

thin deposits of SB may form. The architecture consists of numerous thin, tabular, intersecting sheets. Erosion surfaces, including cutbanks, are common, but may be difficult to identify where gravel units rest on each other.

Model 3 occurs in larger gravel-bed streams, such as trunk rivers, and in some large alluvial fans. The valley contains three or four distinct topographic levels, with the higher levels covered by sparse to dense vegetation. The lowest level is that of the active channel and is similar in all respects to that of model 2. Higher levels are active only during high stage and characteristically accumulate deposits of SB. A floodplain may or may not form a significant part of the system, depending on valley width and channel stability (Fig. 9). Lateral migration of channels, as for example by distributary shifting on alluvial fans, causes superimposition of successively higher terace levels, and the generation of upward-fining sequences (Williams and Rust, 1969; Rust, 1972). These may be thicker than the depth of the channel if they are developed by distributary migration on a rapidly subsiding fan.

Model 4 typifies gravelly rivers of high sinuosity. Typically there is one main, active channel with bars and islands and occasional subsidiary channels. The latter commonly are initiated as chute channels. Sedimentation occurs on large, flat-topped point bar and side bar complexes. These commonly show a downstream decrease in grain size, with gravel sheets, lobes or foreset bars at the head, and sand dunes or sand waves at the tail. Lateral accretion of these bar complexes is common, and the LA element should be recognizable in large outcrops (Figs. 5, 10).

Model 5 represents the typical "coarse-grained meandering stream", with distinctive, gravel-sand or pebbly-sand point bar complexes (Fig. 11). The accretionary face of the bar is crossed by numerous sandy bedforms, including dunes and sand waves. Meander scars and abandoned channels are common in the floodplain. Fining-upward cycles may or may not be developed, depending on meander sinuosity and flow patterns around the bend (Jackson, 1976a). The upper South Platte River (Crowley, 1983) and the Amite River (McGowen and Garner, 1970) are typical modern examples.

Model 6 illustrates the classic sandy meandering stream (Fig. 12). The point bar accretionary face usually is of simpler geometry, with fewer, smaller scale bedforms than in model 5. Accordingly, well developed epsilon crossbedding is to be expected in cross section. Meander scars, abandoned channels and crevasse splays are common.

Model 7 illustrates a highly sinuous, suspended load stream (Fig. 13). The overall geometry is similar to that of model 6, but differs in detail because of the finer grained sediment load (fine sand, silt, mud). Point bar accretion surfaces dip steeply (up to 25°), and have a simple geometry, typically planar or with banks or benches indicating downstream flow separation and the development of incipient scroll bars (Nanson, 1980). Ripple marks are typically the most abundant flow regime bedform present. Gravel lags and crossbedded medium to coarse sands may occur at the base of the point bar. E.H. Koster (pers. com., 1985) reports that many examples of this model may be estuarine in origin.

Models 5, 6 and 7 generate sheet sand bodies, in the terminology of Friend (1983). Their overall composition is that of major CH elements, but many are composed largely of LA deposits, with subordinate SB (GB in model

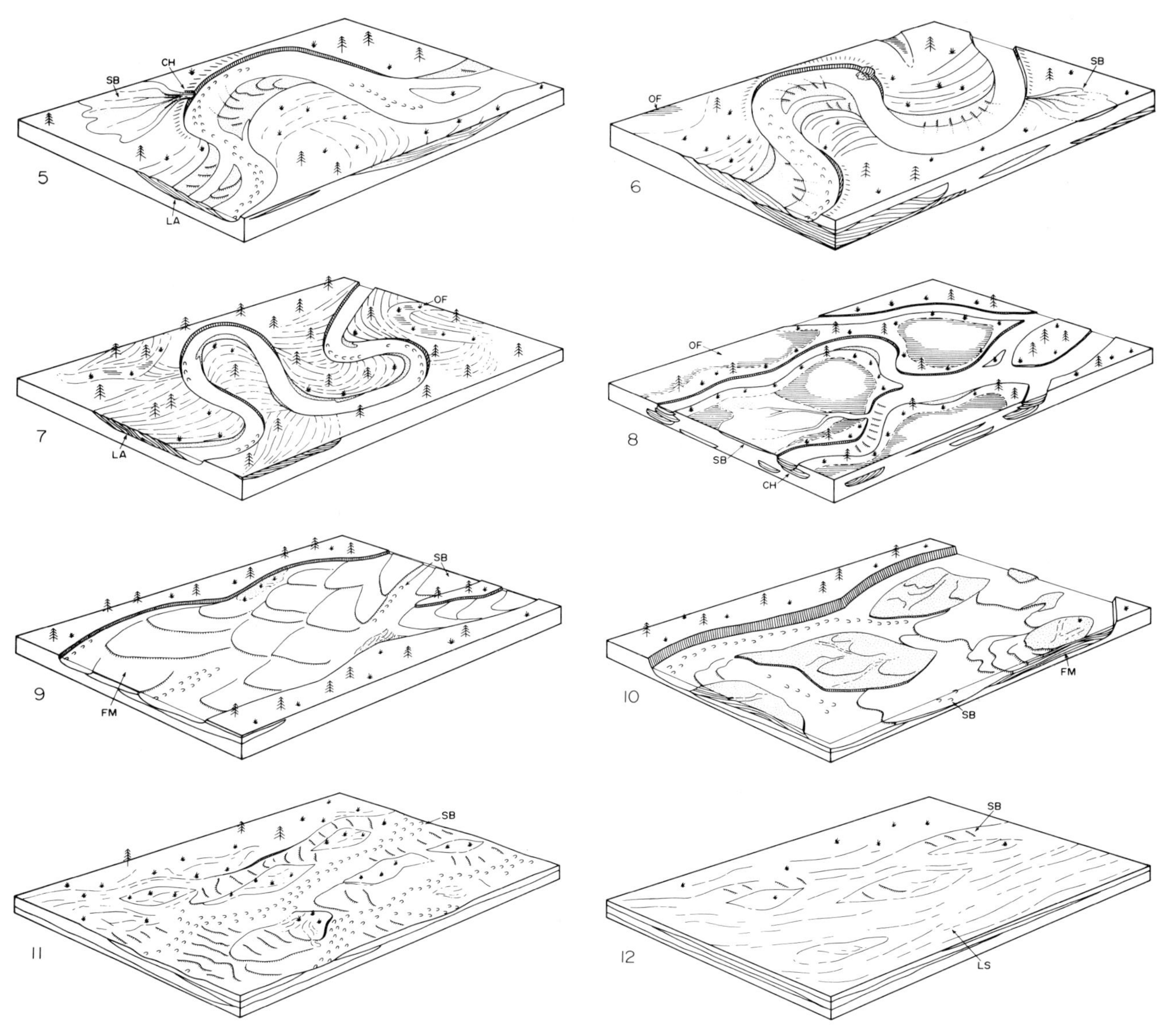

Fig. 10. Model 4: gravelly, high sinuosity river.

Fig. 11. Model 5: sand- and pebbly sand-bed "coarse-grained meandering" river.

Fig. 12. Model 6: the classic sandy, mixed-load meandering river.

Fig. 13. Model 7: muddy, fine grained meandering river.

Fig. 14. Model 8: low to high sinuosity, stable, anastomosed channel system.

Fig. 15. Model 9: low sinuosity river with linguoid bars and "Platte-type macroforms" (Crowley, 1983).

Fig. 16. Model 10: relatively deep, low sinuosity river with foreset macroforms (sand flats or shoals) and isolated linguoid and transverse bars.

Fig. 17. Model 11: distal braidplain, typically ephemeral.

Fig. 18. Model 12: sheetflood fluvial plain subject to highly flashy discharge.

5), OF in abandoned channels, and minor CH deposits representing chute channels. The large scale architecture of the succession is controlled by channel switching and stacking behaviour. As shown by Allen (1974, 1978) and Bridge and Leeder (1979) this depends largely on basin subsidence rates and rate of source area uplift.

Model 8 illustrates the low to high sinuosity, stable, branching channels of the well-known anastomosed fluvial model (Smith, 1983). Channels deposit ribbon sand bodies (cf. Friend et al., 1979; Friend, 1983) composed of element SB. Lateral accretion deposits form a rare to minor component. Floodplains may be wetlands (Smith, 1983) or areas of desiccation and calcrete development (Rust, 1981; Rust and Legun, 1983). Crevasse channels and crevasse splays are common (Fig. 14).

Model 9 encompasses broad, shallow, low sinuosity streams carrying an abundant sand bedload. The channel is filled with fields of large linguoid bedforms, with the deeper channels occupied by trains of dunes (Fig. 15). The linguoid bedforms (commonly termed linguoid bars) have sinuous to lobate avalanche faces and stoss surfaces dipping gently upstream that are covered by ripples or megaripples. These linguoid bars are commonly (but not always) organized into macroforms advancing obliquely down the channel. Successive avalanche faces of the constituent bars terminate at a major avalanche face with a relatively straight crest arranged typically at about 30° to the channel trend (Crowley, 1983).

Preservation of a complete macroform deposit may result in a coarsening-upward sequence (Crowley, 1983). Partial sequences consist of numerous superimposed sets of Sp (the Platte model of Miall, 1977). Other lithofacies, including St (channel floor dunes), Sr (bar top ripples) and

Gm (channel lags) are rare. In this type of river the greater part of the channel typically is undergoing active bedform migration. Stable, vegetated islands are rare except for erosional floodplain remnants, unlike the constructed sandflat complexes of the next model. The typical architecture of this model consists of tabular sheets of SB.

Model 10 is characterized by a much greater differentiation between channel, bar and bar-top facies than model 9, possibly because of greater channel depth or stage fluctuation (Fig. 16). Most of the constructional activity takes place within large macroforms that have variously been termed sand flats (Cant and Walker, 1978), sand shoals (Allen, 1983) or compound bars (Miall, 1977, 1981). The architecture of two ancient examples has been well described by Haszeldine (1983a, b) and Kirk (1983), and is illustrated in Figure 4.

Many deposits of this type show fining-upward cycles (Cant and Walker, 1978; South Saskatchewan model of Miall, 1978), recording the superimposition of channel, bar and bar-top deposits. Such cycles cannot be thicker than the depth of the channel, whereas cycles tens of metres in thickness may record long-term braidplain aggradation and channel switching (as on alluvial fans), or may be tectonic in origin. Similar thick cyclic sequences have been recorded in gravel-sand deposits of model 3 (Miall, 1984) and probably have a similar origin.

Model 11 typifies distal braidplains, particularly in arid regions where ephemeral run-off forms a network of shallow, interlacing, poorly defined channels (Williams, 1971). The deposits are dominated by sheets, lenses and wedges of SB, with rare overbank deposits (Fig. 17). Flood cycles up to about 3 m thick are common (eg. Miall and Gibling, 1978).

Model 12 is a variation on model 11, in which the deposits are dominated by element LS (principally lithofacies Sh), indicating extremely flashy discharge. Channels are poorly defined, so that tabular sand body geometries are typical. Overbank deposits are rare to absent (Fig. 18).

CONCLUSIONS

The virtue of the method of architectural-element analysis is that it reverts to the purely descriptive. A framework for carrying out such an analysis is provided here, which frees the sedimentologist from rigid adherence to any preconceived model. Once the elements have been pulled apart, analyzed, and their method of superposition deduced, a local geomorphologic model can be constructed. This approach should free the sedimentologist from such misleading assumptions as:

- lateral accretion deposits represent simple point bars
- lateral accretion deposits are indicators of meandering streams
- meandering streams are not characteristically gravel-rich
- braided streams are always of low sinuosity
- braided streams are rarely associated with accumulations of thick overbank fines
- anastomosed channels are highly sinuous
- meandering channels do not have mid-channel braids or islands.

With a purely empirical geomorphic model in hand the geologist can turn to such questions as the nature of the sedimentary controls responsible for the channel pattern under study, and the relative importance of tectonics and climate change in the construction of the local stratigraphy.

REFERENCES

Allen, J.R.L., 1963: The classification of cross-stratified units with notes on their origin; Sedimentology, v. 2, p. 93-114.

Allen, J.R.L., 1965: The sedimentation and palaeogeography of the Old Red Sandstone of Anglesey, North Wales; Yorks. Geol. Soc. Proc., v. 35, p. 139-185.

Allen, J.R.L., 1968: Current ripples; North Holland Pub. Co., Amsterdam, 433 p.

Allen, J.R.L., 1970: Studies in fluviatile sedimentation: a comparison of fining-upwards cyclothems with special reference to coarse-member composition and interpretation; J. Sediment. Petrol., v. 40, p. 298-323.

Allen, J.R.L., 1974: Studies in fluviatile sedimentation: implications of pedogenic carbonate units, Lower Old Red Sandstone, Anglo-Welsh outcrop, Geol. Jour., v. 9, p. 181-208.

Allen, J.R.L., 1978: Studies in fluviatile sedimentation: an exploratory quantitative model for the architecture of avulsion controlled alluvial suites; Sediment. Geol., v. 21, p. 129-147.

Allen, J.R.L., 1983: Studies in fluviatile sedimentation: bars bar-complexes and sandstone sheets (low-sinuosity braided streams) in the Brownstones (L. Devonian), Welsh Borders; Sediment. Geol., v. 33, p. 237-293.

Allen, J.R.L., and Williams, B.P.J., 1981: Sedimentology and stratigraphy of the Townsend Tuff Bed (Lower Old Red Sandstone) in South Wales and the Welsh Borders; J. Geol. Soc. London, v. 138, p. 15-29.

Allen, J.R.L., and Williams, B.P.J., 1982: The architecture of an alluvial suite: rocks between the Townsend Tuff and Pickard Bay Tuff Beds

(Early Devonian), Southwest Wales; Phil. Trans. Roy. Soc. London, Series B, v. 297, p. 51-89.

Beutner, E.C., Flueckinger, L.A. and Gard, T.M., 1967: Bedding geometry in a Pennsylvanian channel sandstone; Geol. Soc. Amer. Bull., v. 78, p. 911-916.

Blakey, R.C. and Gubitosa, R., 1984: Controls of sandstone body geometry and architecture in the Chinle Formation (Upper Triassic), Colorado Plateau; Sediment. Geol., v. 38, p. 51-86.

Blodgett, R.H., and Stanley, K.O., 1980: Stratification, bedforms, and discharge relations of the Platte braided river system, Nebraska; J. Sediment. Petrol., v. 50, p. 139-148.

Bluck, B.J., 1971: Sedimentation in the meandering River Endrick; Scot. Jour. Geol., v. 7, p. 93-138.

Bluck, B.J., 1976: Sedimentation in some Scottish rivers of low sinuosity; Roy. Soc. Edinburgh, Trans., v. 69, p. 425-456.

Bluck, B.J., 1979: Structure of coarse grained braided stream alluvium, Trans. Roy. Soc. Edinburgh, v. 70, p. 181-221.

Bluck, B.J., 1980: Structure, generation and preservation of upward fining braided stream cycles in the Old Red Sandstone of Scotland, Trans. Roy. Soc. Edin.; Earth Sciences, v. 71, p. 29-46.

Boothroyd, J.C. and Nummedal D., 1978: Proglacial braided outwash: a model for humid alluvial fan deposits; in A.D. Miall, Ed., Fluvial sedimentology; Can. Soc. Petrol. Geol. Mem. 5, p. 641-668.

Brady, R.H., 1984: Neogene stratigraphy of the Avawatz Mountains between the Garlock and Death Valley fault zones, southern Death Valley, California: implications as to Late Cenozoic tectonism; Sediment. Geol., v. 38, p. 127-158.

Bridge, J.S. and Jarvis, J., 1976: Flow and sedimentary processes in the meandering River South Esk, Glen Clova, Scotland, Earth Surface Processes, v. 1, p. 303-336.

Bridge, J.S. and Leeder, M.R., 1979: A simulation model of alluvial stratigraphy; Sedimentology, v. 26, p. 617-644.

Buck, S.G., 1983: The Saaiplaas Quartzite Member: a braided system of gold and uranium bearing channel placers within the Proterozoic Witwatersrand Supergroup of South Africa; in J.D. Collinson and J. Lewin, eds., Modern and ancient fluvial sediments, Internat. Assoc. Sediment. Spec. Pub. 6, p. 549-562.

Bull, W.B. 1977. The alluvial fan environment, Progr. Phys. Geogr., v. 1, p. 222-270.

Campbell, C.V., 1976: Reservoir geometry of a fluvial sheet sandstone, Am. Assoc. Petrol. Geol. Bull., v. 60, p. 1009-1020.

Cant, D.J. and Walker, R.G., 1978: Fluvial processes and facies sequences in the sandy braided South Saskatchewan River, Canada; Sedimentology, v. 25, p. 625-648.

Collinson, J.D., 1970: Bedforms of the Tana River, Norway; Geogr. Annaler, v. 52A, p. 31-55.

Crowley, K.D., 1983: Large-scale bed configurations (macroforms), Platte River Basin, Colorado and Nebraska: Primary structures and formative processes; Geol. Soc. Am. Bull., v. 94, p. 117-133.

Ethridge, F.G., Jackson, T.J. and Youngberg, A.D., 1981: Floodbasin sequence of a fine-grained meander belt subsystem: the coal-bearing Lower Wasatch and Upper Fort Union Formations, southern Powder River Basin, Wyoming; in F.G. Ethridge and R.M. Flores, eds., Recent and ancient nonmarine depositional environments: models for exploration; Soc. Econ. Paleont. Mineral. Spec. Pub. 31, p. 191-209.

Eyles, N., Eyles, C.H. and Miall, A.D., 1984: Lithofacies types and vertical profile models; an alternative approach to the description and environmental interpretation of glacial diamict and diamictite sequences; Sedimentology, v. 30, p. 393-410.

Forbes, D.L., 1983: Morphology and sedimentology of a sinuous gravel-bed channel system: lower Babbage River, Yukon coastal plain, Canada; in J.D. Collinson and J. Lewin, eds., Modern and ancient fluvial sediments, Internat. Assoc. Sediment. Spec. Pub. 6, p. 195-206.

Friend, P.F., 1983: Towards the field classification of alluvial architecture or sequence, in J.D. Collinson and J. Lewin, eds; Modern and ancient fluvial systems, Internat. Assoc. Sedimentologists Spec. Pub. 6.

Friend, P.F., Slater, M.J. and Williams, R.C., 1979: Vertical and lateral building of river sandstone bodies, Ebro Basin, Spain; J. Geol. Soc. London, v. 136, p. 39-46.

Gustavson, T.C., 1978: Bed forms and stratification types of modern gravel meander lobes, Nueces River, Texas; Sedimentology, v. 25, p. 401-426.

Harms, J.C., and Fahnestock, R.K., 1965: Stratification, bed forms, and flow phenomena (with an example from the Rio Grande); in G.V. Middleton, ed., Primary Sedimentary Structures and their Hydrodynamic interpretation, Spec. Pub. 12, p. 84-115.

Harms, J.C., Mackenzie, D.B., and McCubbin, D.G., 1963: Stratification in modern sands of the Red River, Louisiana, J. Geol. v.71, p. 566-580.

Harms, J.C., Southard, J.B., Spearing, D.R. and Walker, R.G., 1975: Depositional environments as interpreted from primary sedimentary structures and stratification sequences; Soc. Econ. Paleont. Mineral. Short Course No. 2, 161 p.

Harms, J.C., Southard, J.B. and Walker, R.G., 1982: Structures and sequences in clastic rocks; Soc. Econ. Paleont. Mineral. Short Course No. 9.

Haszeldine, R.S., 1983a: Fluvial bars reconstructed from a deep, straight channel, Upper Carboniferous coalfield of northeast England; J. Sediment. Petrol. v. 53, p. 1233-1248.

Haszeldine, R.S., 1983b: Descending tabular cross-bed sets and bounding surfaces from a fluvial channel in the Upper Carboniferous coalfield of north-east England, in J.D. Collinson and J. Lewin, Ed., Modern and ancient fluvial systems; Internat. Assoc. Sedimentologists Spec. Pub. 6, p. 449-456.

Hein, F.J. and Walker, R.G., 1977: Bar evolution and development of stratification in the gravelly, braided, Kicking Horse River, British Columbia, Can. J. Earth Sci., v. 14, p. 562-570.

Hooke, R. LeB., 1967: Processes on arid-region alluvial fans; Geol., v. 75, p. 438-460.

Hopkins, J.C., Hermanson, S.W., and Lawton, D.C., 1982: Morphology of channels and channel-sand bodies in the Glauconitic Sandstone Member (Upper Mannville), Little Bow area, Alberta; Bull. Can. Petrol. Geol., v. 30, p. 274-285.

Jackson, R.G. II., 1975: Hierarchial attributes and a unifying model of bed forms composed of cohesionless material and produced by shearing flow, Geol. Soc. Am. Bull., v. 86, p. 1523-1533.

Jackson, R.G. II., 1976a: Depositional model of point bars in the lower Wabash River, J. Sediment. Petrol., v. 46, p. 579-594.

Jackson, R.G., 1976b: Large scale ripples of the lower Wabash River; Sedimentology, v. 23, p. 593-632.

Jackson, R.G. II, 1978: Preliminary evaluation of lithofacies models for meandering alluvial streams; in A.D. Miall, ed., Fluvial Sedimentology, Can. Soc. Petrol. Geol., Mem. 5, p. 543-576.

Kirk, M., 1983: Bar developments in a fluvial sandstone (Westphalian "A"), Scotland; Sedimentology, v. 30, p. 727-742.

Leeder, M.R., 1973: Fluviatile fining-upward cycles and the magnitude of paleochannels; Geol. Mag., v. 110, p. 265-276.

Long, D.G.F., 1978: Proterozoic stream deposits: some problems of recognition and interpretation of ancient sandy fluvial systems; in A.D. Miall, Ed., Fluvial Sedimentology, Can. Soc. Petrol. Geol. Mem. 5, p. 313-342.

Massari, F., 1983: Tabular cross-bedding in Messinian fluvial channel conglomerates, southern Alps, Italy; in J.D. Collinson and J. Lewin, eds., Modern and ancient fluvial systems; Internat. Assoc. Sediment. Spec. Pub. 6, p. 287-300.

McGowen, J.H. and Garner, L.E., 1970: Physiographic features and stratification types of coarse-grained point bars; modern and ancient examples; Sedimentology, v. 14, p. 77-112.

McKee, E.D., Crosby, E.J., and Berryhill, H.L., 1967: Flood deposits, Bijou Creek, Colorado; J. Sediment. Petrol.,v. 37, p. 829-851.

McKee, E.D. and Weir, G.W., 1953: Terminology for stratification and cross-stratification in sedimentary rocks; Geol. Soc. Amer. Bull., v. 64, p. 381-389.

Miall, A.D., 1973: Markov chain analysis applied to an ancient alluvial plan succession; Sedimentology, v. 20, p. 347-364.

Miall, A.D., 1977: A review of the braided river depositional environment; Earth Sci. Reviews, v. 13, p. 1-62.

Miall, A.D., 1978: Lithofacies types and vertical profile models in braided river deposits: a summary; in A.D. Miall, ed., Fluvial sedimentology, Can. Soc. Petrol. Geol. Mem. 5, p. 597-604.

Miall, A.D., 1980: Cyclicity and the facies model concept in fluvial deposits; Bull. Can. Petrol. Geol., v. 28, p. 59-80.

Miall, A.D., 1981: Analysis of fluvial depositional systems; Am. Assoc. Petrol. Geol. Education Course Note Series 20, 75p.

Miall, A.D., 1984: Variations in fluvial style in the Lower Cenozoic synorogenic sediments of the Canadian Arctic Islands; Sediment. Geol., v. 38, p. 499-523.

Miall, A.D. and Gibling, M.R., 1978: The Siluro-Devonian clastic wedge of Somerset Island, Arctic Canada, and some regional paleogeographic implications; Sedimentary Geology, v. 21, p. 85-127.

Mossop, G.D. and Flach, P.D., 1983: Deep channel sedimentation in the Lower Cretaceous McMurray Formation, Athabasca Oil Sands, Alberta; Sedimentology, v. 30, p. 493-509.

Nanson, G.C., 1980: Point bar and floodplain formation of the meandering Beatton River, northeastern British Columbia, Canada; Sedimentology, v. 27, p. 3-30.

Nanson, G. and Page, K., 1983: Lateral accretion of fine-grained concave benches on meandering rivers; in J.D. Collinson and J. Lewin, eds., Modern and ancient fluvial systems, Internat. Assoc. Sediment. Spec. Pub. 6, p. 133-144.

Nemec, W. and Muszynski, A., 1982: Volcaniclastic alluvial aprons in the Tertiary of Sofia district (Bulgaria); Annales Soc. Geol. Poloniae, v. 52, p. 239-303.

Ori, G.G., 1979: Barre di meandre nelle alluvioni ghiaiose del fiume Reno (Bologna); Bull. Soc. Geol. Italia, v. 98, p. 35-54.

Parkash, B., Awasthi, A.K. and Gohain, K., 1983: Lithofacies of the Markanda terminal fan, Kurukshetra district, Haryana, India; in J.D. Collinson and J. Lewin, eds., Modern and ancient fluvial systems, Internat. Assoc. Sediment. Spec. Pub. 6, p. 337-344.

Puigdefabregas, C., 1973: Miocene point-bar deposits in the Ebro Basin, northern Spain, Sedimentology, v. 20, p. 133-144.

Putnam, P.E., 1982a: Fluvial channel sandstones within Upper mannville (Albian) of Lloydminster area, Canada - geometry, petrography, and paleogeographic implications; Am. Assoc. Petrol. Geol. Bull., v. 66, p. 436-459.

Putnam, P.E., 1982b: Aspects of the petroleum geology of the Lloydminster heavy oil fields, Alberta and Saskatchewan; Bull. Can. Petrol. Geol., v. 30, p. 81-111.

Ramos, A., and Sopena, A., 1983: Gravel bars in low-sinuosity streams (Permian and Triassic, central Spain); in J.D. Collinson and J. Lewin, eds., Modern and ancient fluvial sediments; Internat. Assoc. Sediment. Spec. Pub. 6, p. 301-213.

Rust, B.R., 1972: Structure and process in a braided river; Sedimentology, v. 18, p. 221-246.

Rust, B.R., 1978a: A classification of alluvial channel systems, in A.D. Miall, ed., Fluvial Sedimentology, Can. Soc. Petrol. Geol. Mem. 5, p. 187-198.

Rust, B.R., 1978b: Depositional models for braided alluvium; in A.D. Miall, ed., fluvial Sedimentology, Can. Soc. Petrol. Geol. Mem. 5, p. 605-625.

Rust, B.R., 1981: Sedimentation in an arid-zone anastomosing fluvial system: Cooper's Creek, Central Australia; J. Sediment. Petrol., v. 51, p. 745-755.

Rust, B.R. and Legun, A.S., 1983: Modern anastomosing fluvial deposits in arid central Australia, and a Carboniferous analogue in New Brunswick, Canada; in J.D. Collinson and J. Lewin, eds., Modern and ancient fluvial sediments. Internat. Assoc. Sediment. Spec. Pub. 6, p. 385-392.

Schumm, S.A., 1963: A tentative classification of alluvial river channels, U.S. Geol. Surv. Circ. 477.

Schumm, S.A., 1968: Speculations concerning paleohydrologic controls of terestrial sedimentation; Geol. Amer. Bull., v. 79, p. 1573-1588.

Schwartz, D.E., 1978: Hydrology and current orientation analysis of a braided to meandering transition; the Red River in Oklahoma and Texas, U.S.A., in A.D. Miall, ed., Fluvial sedimentology; Can. Soc. Petrol. Geol. Mem. 5, p. 231-256.

Schultz, A.W., 1984: Subaerial debris-flow deposition in the Upper Paleozoic Cutler Formation, Western Colorado, J. Sediment. Petrol., v. 54, p. 749-772.

Smith, D.G., 1976: Effect of vegetation on lateral migration of anastomosed channels of a glacial meltwater river; Geol. Soc. Am. Bull., v. 87, p. 857-860.

Smith, D.G., 1983: Anastomosed fluvial deposits; modern examples from Western Canada; in J.D. Collinson and J. Lewin, eds., Modern and ancient fluvial systems; Internat. Assoc. Sediment. Spec. Pub. 6, p. 155-168.

Smith, N.D., 1970: The braided stream depositional environment: comparison of the Platte River with some Silurian clastic rocks, north central Appalachians; Geol. Soc. Amer. Bull., v. 81, p. 2993-3014.

Smith, N.D., 1971: Transverse bars and braiding in the lower Platte River, Nebraska; Geol. Soc. Am. Bull., v. 82, p. 3407-3420.

Smith, N.D., 1972: Some sedimentological aspects of planar cross-stratification in a sandy braided river; J. Sediment. Petrol., v. 42, p. 624-634.

Smith, N.D., 1974: Sedimentology and bar formation in the upper Kicking Horse River, a braided outwash stream, J. Geol., v. 82, p. 205-224.

Sneh, A., 1983: Desert stream sequences in the Sinai Peninsula; J. Sediment. Petrol., v. 53, p. 1271-1280.

Southard, J.B., 1971: Representation of bed configurations in depth-velocity-size diagrams; J. Sediment. Petrol., v. 41, p. 903-915.

Tunbridge, I.P., 1981: Sandy high-energy flood sedimentation - some criteria for recognition, with an example from the Devonian of S.W. England; Sediment. Geol., v. 28, p. 79-96.

Vessell, R.K., and Davies, D.K., 1981: Nonmarine sedimentation in an active fore arc basin; in F.G. Ethridge and R.M. Flores, eds., Recent and ancient nonmarine depositional environments: models for exploration; Soc. Econ. Paleont. Mineral. Spec. Pub. 31, p. 31-45.

Vos, R.G. and Tankard, A.J., 1981: Braided fluvial sedimentation in the Lower Paleozoic Cape Basin, South Africa; Sediment. Geol., v. 29, p. 171-193.

Wasson, R.J., 1977: Last-glacial alluvial fan sedimentation in the Lower Derwent Valley, Tasmania; Sediment. Geol., v. 24, p. 781-800.

Wightman, D.M., Tilley, B.J. and Last, W.M., 1981: Stratigraphic traps in channel sandstones in the upper Mannville (Albian) of east-central Alberta: Discussion; Bull. Can. Petrol. Geol., v. 29, p. 622-625.

Williams, G.E., 1971: Flood deposits of the sand-bed ephemeral streams of central Australia, Sedimentology, v. 17, p. 1-40.

Williams, P.F. and Rust, B.R., 1969: The sedimentology of a braided river; J. Sediment. Petrol., v. 39, p. 649-679.

CHAPTER 4

MULTIPLE-CHANNEL BEDLOAD RIVERS

Andrew D. Miall

Geology Department

University of Toronto

Toronto, Ontario M5S 1A1 Canada

INTRODUCTION

Rivers with an abundant gravel or sand sediment load were classified as bedload rivers by Schumm (1963). They are usually described by the term "braided", but in its common useage this term is misleading, because it has come to imply only rivers of low sinuosity (less than 1.5, using the classification of Rust, 1978a). However, bedload rivers with higher sinuosities are relatively common. They may have stable islands and branching channels, and therefore do not fall conveniently into either the braided or meandering categories. Examples are described by Bluck (1971), Gustavson (1978) and Forbes (1983). They are encompassed by architectural model 4 of Miall (this volume) and will be described briefly below.

The terms alluvial fan and fan delta are widely used in the geologic literature. They carry specific geomorphic implications about the dispersal of the water and sediment in a series of distributaries, but these characteristics often cannot be demonstrated from the ancient record. Lithofacies assemblages typical of gravelly alluvial fans, such as the Trollheim and Scott types of Miall (1978), corresponding closely to architectural models 1 and 2 of Miall (this volume), are found both in alluvial fan and braidplain settings. The terms alluvial fan and fan delta should therefore be used with caution when describing the ancient record. The sedimentology of these environments is dealt with by Ethridge (this volume) and Galloway (this volume), and will therefore not be covered in this chapter.

The remaining "braided" models of Miall (1978), comprising architectural models 3, 4, 9 10, 11 and 12 (Miall this volume), are the main subject of this chapter (table 10). Reference is made throughout to the lithofacies classification scheme of Miall (1977, 1978; see Miall, this volume, Table 1).

THE MODELS

As emphasized elsewhere (Miall, this volume), although fluvial sediments can be categorized into a number of models, every possible gradation between any two of the models can be expected to occur in nature. The sedimentologist should therefore avoid too rigid an assignment of any given fluvial deposit to one or other of the models. Numbering of the models follows that of the earlier chapter by this writer.

Model 3

This corresponds to the Donjek model of Miall (1977), which was based on descriptions of the middle reaches of the Donjek River, Yukon, by Williams and Rust (1969).

The fluvial style consists of medium to large-scale gravel-bed streams with well defined, active channels and elevated, partially to entirely inactive channel reaches. Topographic relief across the system may be in the order of 3-7 m. The coarsest sediments occur in the deeper channels, and consist of various lithofacies comprising architectural element GB: gravel bars and bedforms. Partially inactive tracts above the deep channels receive sand and gravel during flood stages, with the dominant sediments comprising element SG: sandy bedforms. The highest parts of the system may be covered by dense vegetation which traps fine sediment from infiltrated flood waters. Accumulations of element OF: overbank fines, are the result, though they may or may not form a significant part of the

Name	Environmental setting	Main facies	Minor facies
Trollheim type (G_I)	proximal rivers (predominantly alluvial fans) subject to debris flows	*Gms, Gm*	*St, Sp, Fl, Fm*
Scott type (G_{II})	proximal rivers (including alluvial fans) with stream flows	*Gm*	*Gp, Gt, Sp, St, Sr, Fl, Fm*
Donjek type (G_{III})	distal gravelly rivers (cyclic deposits)	*Gm, Gt, St*	*Gp, Sh, Sr, Sp, Fl, Fm*
South Saskatchewan type (S_{II})	sandy braided rivers (cyclic deposits)	*St*	*Sp, Se, Sr, Sh, Ss, Sl, Gm, Fl, Fm*
Platte type (S_{II})	sandy braided rivers (virtually non cyclic)	*St, Sp*	*Sh, Sr, Ss, Gm, Fl, Fm*
Bijou Creek type (S_I)	Ephemeral or perennial rivers subject to flash floods	*Sh, Sl*	*Sp, Sr*

Modified from Miall (1977) and Rust (

Table 1. Lithofacies assemblages and facies models for multiple-channel bedload rivers (Miall, 1978). These named models have now been superseded by the more flexible architectural models described here and in an earlier chapter by the writer.

system, depending on valley width, channel stability and the relative rates of channel migration versus subsidence.

The lithofacies assemblage characteristic of this model is shown in Figure 1. Note that it contains fining-upward cycles, that develop as a result of lateral shifts in the various subenvironments. These were first noted by Williams and Rust (1969). A simplified model for the environment is given in Figure 2, and an example is illustrated in Figure 3.

Ancient examples have been described by Miall (1970), Steel (1974), Minter (1978) and Rust (1978b), and additional examples are given by Rust and Koster (1984).

The Scott, Donjek and South Saskatchewan assemblages may occur in a gradational proximal-distal relationship on the same alluvial system (eg. Brady, 1984). It is suggested that the following numerical limits of gravel content (percent of cumulative thickness in a secion) be used to distinguish the three types: Scott >90%, Donjek 10-90%, South Saskatchewan <10%.

Model 4

A category of fluvial style that is receiving increasing attention is the gravel bed river showing high sinuosity and well developed point bars. A detailed description of an example of this type, the River Endrick, Scotland, was described by Bluck (1971) some years ago, but never became established as a standard facies model. More recent descriptions of modern rivers in this category are those of the Nueces River, Texas (Gustavson, 1978) and the lower Babbage River, Yukon (Forbes, 1983).

In vertical profile it would be difficult to distinguish the deposits of this type of river from those of model 3. The main distinguishing characteristic is the presence of gravel lateral accretion deposits (element LA), which could not be recognized in confined outcrops or core.

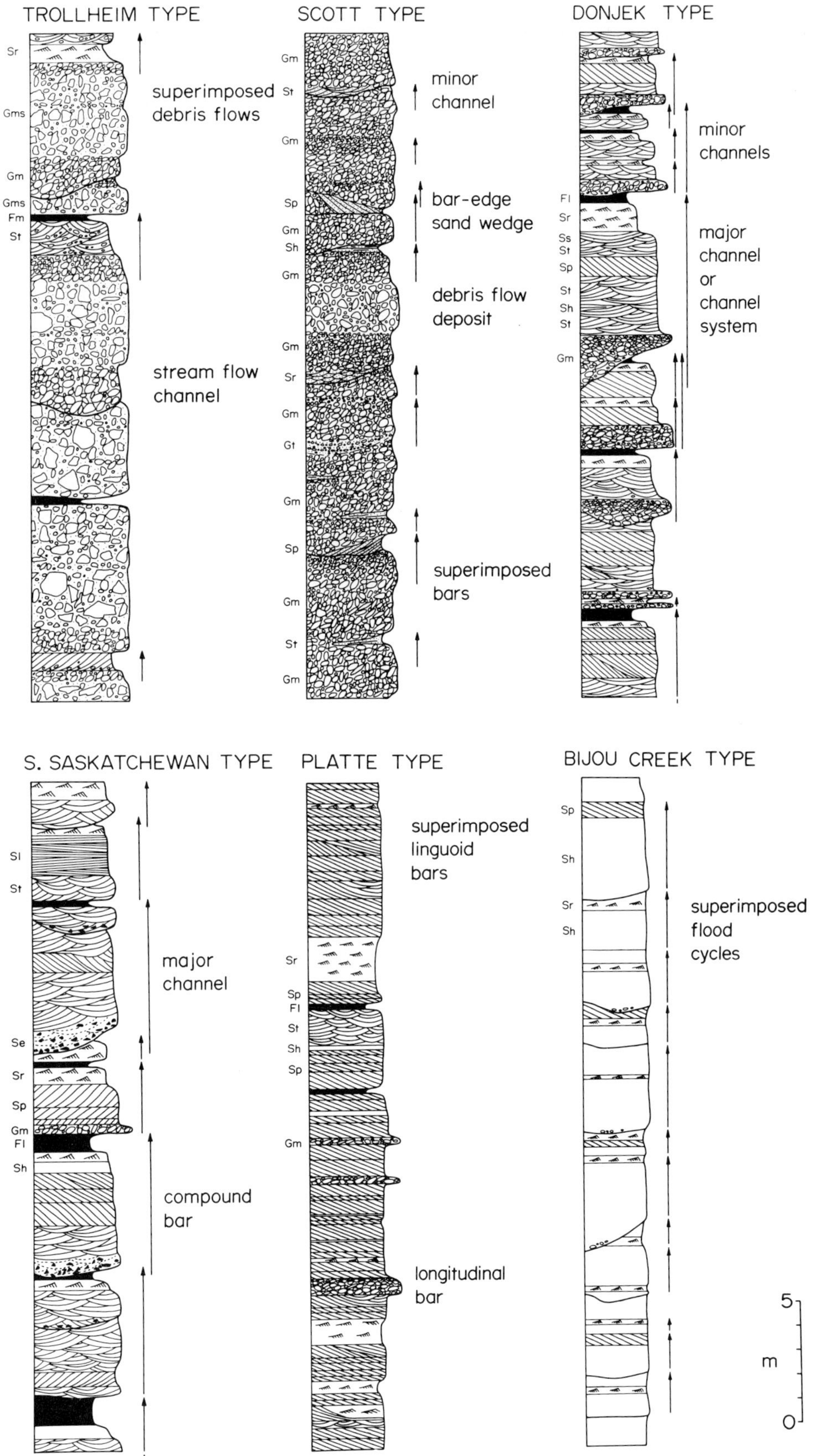

Fig. 1. The six vertical profile models of Miall (1978). The Trollheim and Scott models are not dealt with in this chapter.

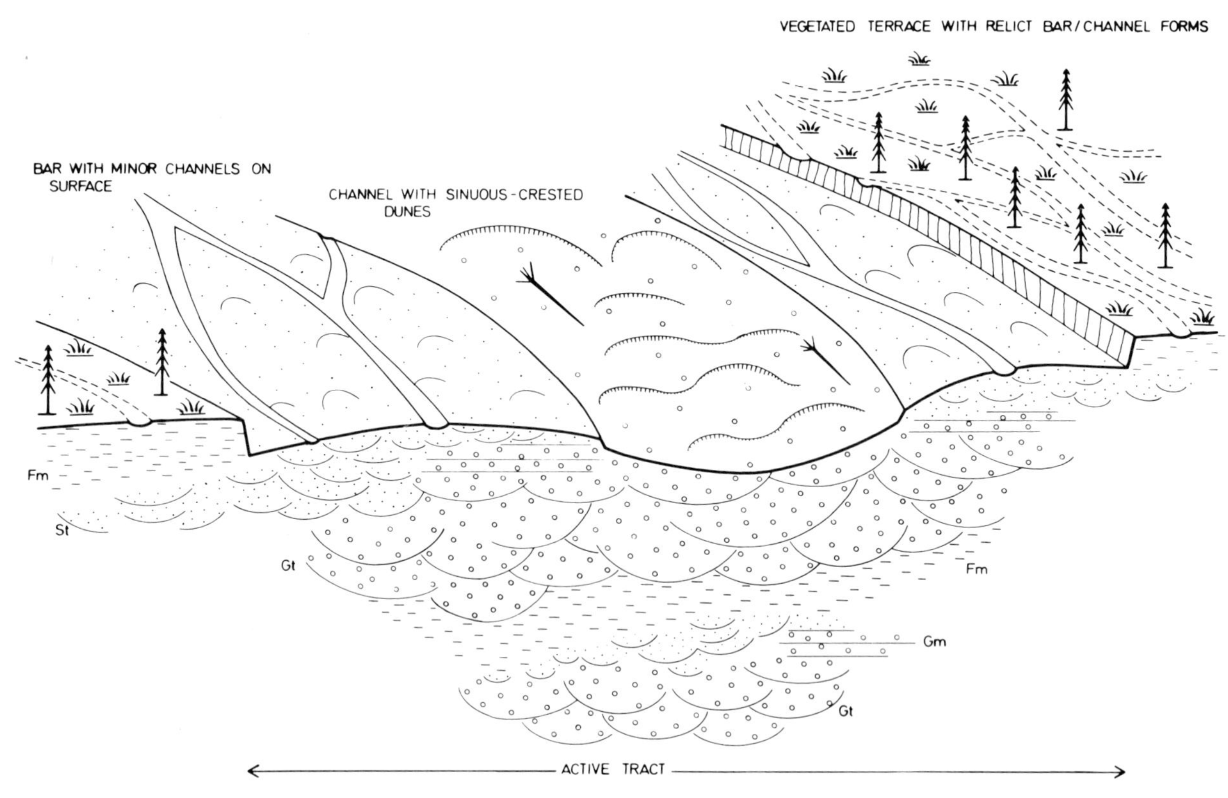

Fig. 2 Facies model developed for the Upper Member of Cannes de Roche Formation, showing a typical Donjek-type fluvial style (Rust and Koster, 1984).

Fig. 3. An ancient example of a Donjek-type fluvial succession. Peel Sound Formation, northern Prince of Wales Island, Arctic Canada (Miall, 1970).

Ancient examples have been described by Ori (1979, 1982), Arche (1983), and Ramos and Sopena (1983). Ori (1979) illustrates a large, gently dipping point bar consisting of lithofacies Gm, overlain by a chute channel filled with lithofacies Gt (see Miall, this volume, Fig. 5A).

Model 9

This model encompasses broad, shallow rivers lacking well defined topographic differention between active and inactive areas. The Platte River (Smith, 1970, 1971, 1972) is a typical example.

The sediments are dominated by element SB: Sandy bedforms, including linguoid or transverse sandy foreset bars, or sand waves, giving rise to superimposed sets of lithofacies Sp. Lithofacies St is sparse, reflecting the rarity of deep channels. Bar top sands and silts (lithofacies Sr, Sh, Fl, Fm) may occur, as in the South Saskatchewan type. Few cycles are developed. A typical profile is illustrated in Figure 1, and an ancient example is show in Fig. 4.

As noted in an earlier chapter the linguoid bars may be organized into macroforms. Crowley (1983) re-examined the Platte River and defined a type of small scale coarsening-upward cycle that may result from complete preservation of a macroform. However, internal scour would seem to be capable of removing the upper part of the cycles and, in fact, none have yet been defined from the ancient record, to the writer's knowledge.

Model 10

Sedimentation is characterised by the development of large sandy macroforms, a discussion of which is given elsewhere (Miall, this volume). These may show a complex internal structure that cannot be deduced from a few widely spaced outcrops or cores.

A well described modern example is the South Saskatchewan river (Cant

Fig. 4. An ancient example of a Platte-type fluvial succession. Isachsen Formation, Banks Island, Arctic Canada (Miall, 1976).

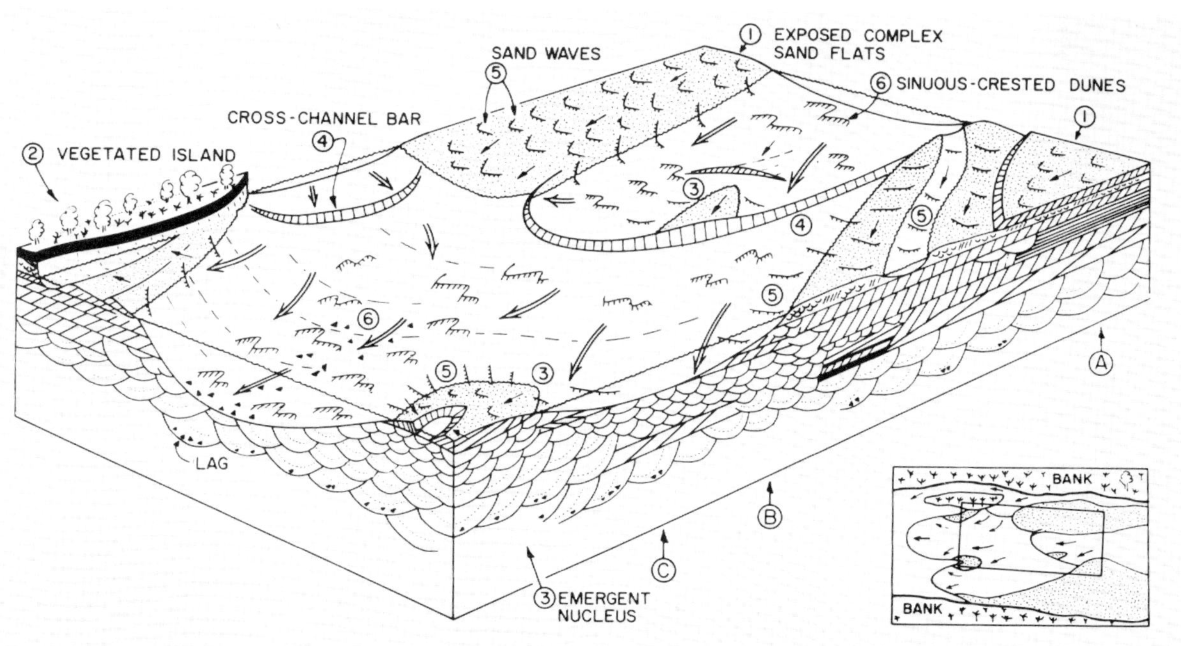

Fig. 5. Block diagram model for the South Saskatchewan River (Walker and Cant, 1984).

The sediments are dominated by element SB: Sandy bedforms, including linguoid or transverse sandy foreset bars, or sand waves, giving rise to superimposed sets of lithofacies Sp. Lithofacies St is sparse, reflecting the rarity of deep channels. Bar top sands and silts (lithofacies Sr, Sh, Fl, Fm) may occur, as in the South Saskatchewan type. Few cycles are developed. A typical profile is illustrated in Figure 1, and an ancient example is show in Fig. 4.

As noted in an earlier chapter the linguoid bars may be organized into macroforms. Crowley (1983) re-examined the Platte River and defined a type of small scale coarsening-upward cycle that may result from complete preservation of a macroform. However, internal scour would seem to be capable of removing the upper part of the cycles and, in fact, none have yet been defined from the ancient record, to the writer's knowledge.

Model 10

Sedimentation is characterised by the development of large sandy macroforms, a discussion of which is given elsewhere (Miall, this volume). These may show a complex internal structure that cannot be deduced from a few widely spaced outcrops or cores.

A well described modern example is the South Saskatchewan river (Cant and Walker, 1978; Walker and Cant, 1984). A facies model for this river is illustrated in Figure 5 and an ancient example is shown in Figure 6.

Rivers of this type display well differentiated channel, bar and bar-top facies. Channels commonly are floored by lag gravels (lithofacies Gm), above which sand is transported as a coarse bedload. Bedforms in the deeper channels (3m or deeper) tend to be sinuous-crested dunes that give rise to lithofacies St. Sandwaves (lithofacies Sp) are common in shallow channel reaches. Sandy foreset bars of a variety of types migrate downstream. In the South Saskatchewan River (but not necessarily in all sandy braided rivers) large sand flats may develop by the coalescence of small bars around a nucleus. These aggrade upstream and may also grow and Walker, 1978; Walker and Cant, 1984). A facies model for this river is illustrated in Figure 5 and an ancient example is shown in Figure 6.

Rivers of this type display well differentiated channel, bar and bar-top facies. Channels commonly are floored by lag gravels (lithofacies Gm), above which sand is transported as a coarse bedload. Bedforms in the deeper channels (3m or deeper) tend to be sinuous-crested dunes that give rise to lithofacies St. Sandwaves (lithofacies Sp) are common in shallow channel reaches. Sandy foreset bars of a variety of types migrate downstream. In the South Saskatchewan River (but not necessarily in all sandy braided rivers) large sand flats may develop by the coalescence of small bars around a nucleus. These aggrade upstream and may also grow downstream by the development of straight-crested bars at high angles to the channel trend (cross-channel bars). One the bar tops lithofacies Sr and Sh will form during flood submergence, and there may also be fine grained overbank deposits in inactive areas.

Typical cycles are illustrated in Figure 1. These are similar to some point bar cycles of meandering rivers. In vertical profile they may be distinguished by careful paleocurrent work; for example Figure 7 illustrates the divergent current directions of bar and sand flat deposits, a quite different pattern to that expected on a fluvial point bar.

Ancient examples of this assemblage have been described by Cant and Walker (1976), Miall and Gibling (1978), Minter (1978) and Rust (1978b).

Model 11

Sheet-braided deposits form in some distal braidplain environments, particularly in arid regions, where runoff and sedimentation are controlled by flash floods.

In vertical profile these deposits may appear to be superficially similar to those of model 10. However, the flood cycles are in general

Fig. 6. An ancient example of a South-Saskatchewan type of fluvial sequence. Old Red Sandstone (Devonian), northern Scottish coast.

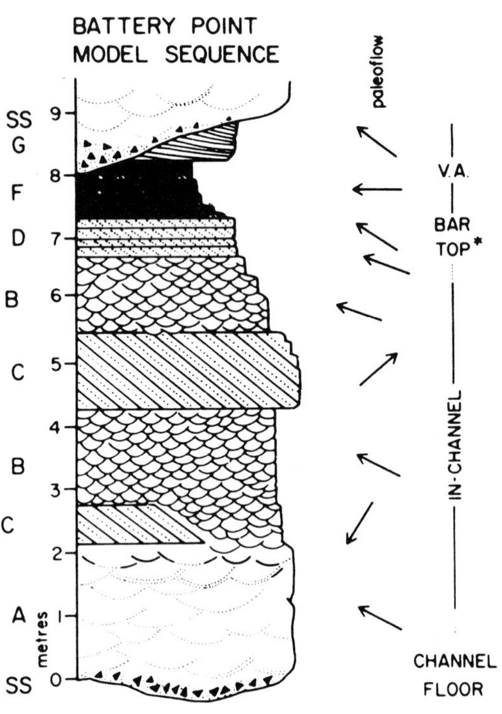

Fig. 7. Current directions in a South Saskatchewan type braided river deposit (north at top). Note consistent WNW pattern of troughs and ripples and divergent pattern of planar crossbed sets (Walker and Cant, 1984).

thinner than the fining-upward cycles of model 10, reflecting the shallower nature of the channels and an origin by rapid, short term aggradation rather than by long term channel migration and macroform accretion. The architecture is sheet-like, and lacks well-defined cutbanks. An ancient example is illustrated in Figure 8, (from Miall and Gibling, 1978). Modern examples are described by Williams (1971).

Model 12

This is a variant of model 11 that occurs in cases of unusually rapid flood runoff, and is dominated by architectural element LS. Under upper flow regime conditions sand is moulded into a plane bed, and may develop parting or streaming lineation on bedding plane surfaces (lithofacies Sh). This exceptionally high flow energy is rare in perennial rivers but is common in ephemeral rivers characterized by infrequent, violent flash floods. Flow may either be channelized or occur as sheet floods. Examples include Bijou Creek, Colorado (McKee et al., 1967) after which the assemblage type illustrated in Figure 1 is named, and parts of the Lake Eyre Basin (Williams, 1971). General descriptions are also given by Picard and High (1973). an ancient example is shown in Figure 9.

Facies models for ephemeral rivers have been discussed by Miall (1977), Rust (1978b), and Tunbridge (1981). They all emphasize the importance of lithofacies Sh in the sedimentary assemblage. Lithofacies Sp, Sr, Fl and Fm may be produced in the waning flood stages, generating thin fining-upward cycles. Thicknesses of sediment in excess of 1.5m may be deposited in a single flood. Erosional surfaces with intraclasts (lithofacies Se) and low angle crossbeds filling scours (lithofacies Sl) may also be important, as in the "Malbaie type" assemblage described by Rust (1978b). Desiccation features may also be important.

Fig. 8. Flood cycles in the Peel Sound Formation, northern Somerset Island, Arctic Canada (Miall and Gibling, 1978).

Fig. 9. An example of a Bijou Creek-type sequence. Eureka Sound Formation, Ellesmere Island, Arctic Canada (Miall, 1984).

References

Arche, A., 1983: Coarse-grained meander lobe deposits in the Jarama River, Madrid, Spain; in J.D. Collinson and J. Lewin, eds., Modern and ancient fluvial sediments, Internat. Assoc. Sediment. Spec. Pub. 6, p. 313-322.

Bluck, B.J., 1971: Sedimentation in the meandering River Endrick; Scot. J. Geol., v. 7, p. 93-138.

Brady, R.H., 1984: Neogene stratigraphy of the Avawatz Mountains between the Garlock and Death Valley fault zones, southern Death Valley, California: implications as to Late Cenozoic tectonism; Sediment. Geol., v. 38, p. 127-158.

Cant, D.J. and Walker, R.G., 1976: Development of a braided-fluvial facies model for the Devonian Battery Point Sandstone, Quebec; Can. J. Earth Sci., v. 13, p. 102-119.

Cant, D.J. and Walker, R.G., 1978: Fluvial proceses and facies sequences in the sandy braided South Saskatchewan River, Canada; Sedimentology, v. 25, p. 625-648.

Crowley, K.D., 1983: Large-scale bed configurations (macroforms), Platte River Basin, Colorado and Nebraska: Primary structures and formative processes; Geol. Soc. Am. Bull., v. 94, p. 117-133.

Forbes, D.L., 1983: Morphology and sedimentology of a sinuous gravel-bed channel system: lower Babbage River, Yukon coastal plain, Canada; in J.D. Collinson and J. Lewin, eds., Modern and ancient fluvial systems, Internat. Assoc. Sediment Spec. Pub. 6, p. 195-206.

Gustavson, T.C., 1978: Bed forms and stratification types of modern gravel meander lobes; Sedimentology, Nueces River, Texas; v. 25, p. 401-426,

McKee, E.D., Crosby, E.J., and Beryhill, H.L., 1967: Flood deposits, Bijou Creek, Colorado; J. Sediment. Petrol., v. 37, p. 829-851.

Miall, A.D., 1970: Continental - marine transition in the Devonian of Prince of Wales Island, Northwest Territories, Can. J. Earth Sci., v. 7, p. 125-144.

Miall, A.D., 1978: Paleocurrent and paleohydrologic analysis of some vertical profiles through a Cretaceous braided stream deposit, Banks Island, Arctic Canada; Sedimentology, v. 23, p. 459-484.

Miall, A.D., 1977: A review of the braided river depositional environment; Earth Sci. Reviews, v. 13, p. 1-62.

Miall, A.D., 1978: Lithofacies types and vertical profile models in braided river deposits: a summary; in A.D. Miall, ed., Fluvial sedimentology, Can. Soc. Petrol. Geol. Mem. 5, p. 597-604.

Miall, A.D., and Gibling, M.R., 1978: The Siluro-Devonian clastic wedge of Somerset Island, Arctic Canada, and some regional paleogeographic implications; Sedimentary Geology, v. 21, p. 85-127.

Minter, W.E.L., 1978: A sedimentological synthesis of placer gold, uranium and pyrite concentrations in Proterozoic Witwatersrand sediments; in A.D. Miall, ed., Fluvial sedimentology; Can. Soc. Petrol. Geol. Mem. 5, p. 801-829.

Ori, G.G., 1979: Barre di meandro nelle alluvioni ghiaiose del fiume reno (Bologna), Boll. Soc. Geol. Italy, v. 98, p. 35-54.

Ori, G.G., 1982: Braided to meandering channel patterns in humid-region alluvial fan deposits, River Reno, Po Plain (northern Italy); Sediment. Geol., v. 31, p. 231-248.

Picard, M.D., and High, L.R., jr., 1973: Sedimentary structures of ephemeral streams, Dev. in Sed., No. 17, Elsevier, Amsterdam, 223 p.

Ramos, A., and Sopena, A., 1983: Gravel bars in low-sinuosity streams (Permian and Triassic, central Spain); in J.D. Collinson and J. Lewin, eds., Modern and ancient fluvial systems, Internat. Assoc. Sediment.

Spec. Pub. 6, p. 301-312.

Rust, B.R., 1978a: A classification of alluvial channel systems; in A.D. Miall, ed., Fluvial sedimentology, Can. Soc. Petrol. Geol., Mem. 5, p. 187-198.

Rust, B.R., 1978b. Depositional models for braided alluvium; in A.D. Miall, ed., Fluvial Sedimentology, Can. Soc. Petrol. Geol. Mem. 5, p. 605-625.

Rust, B.R., and Koster, E.H., 1984: Coarse alluvial deposits; in R.G. Walker, ed., Facies models, second edition, Geoscience Canada Reprint Series no. 1, p. 53-69.

Schumm, S.A., 1963: A tentative classification of alluvial river channels, USGS Circular 477, 10p.

Smith, N.D., 1970: The braided stream depositional environment: Comparison of the Platte River with some Silurian clastic rocks, north central Appalachians; Geol. Soc. Amer. Bull, v. 81, p. 2993-3014.

Smith, N.D., 1971; Transverse bars and braiding in the lower Platte River, Nebraska; Geol. Soc. Am. Bull., v. 82, p. 3407-3420.

Smith, N.D., 1972: Some sedimentological aspects of planar cross-stratification in a sandy braided river; J. Sediment. Petrol., v. 42, p. 624-634.

Tunbridge, I.P., 1981, Sandy high-energy flood sedimentation - some criteria for recognition, with an example from the Devonian of S.W. England, Sediment. Geol., v. 28, p. 79-96.

Steel, R.J., 1974: New Red Sandstone floodplain and piedmont sedimentation in the Hebridean Province, Scotland, J. Sediment, Petrol., v. 44, p. 336-357.

Walker, R.G. and Cant, D.J., 1984: Sandy fluvial systems; in R.G. Walker, ed., Facies models, second edition, Geoscience Canada Reprint Series

#1, p. 71-89.

Williams, G.E., 1971: Flood deposits of the sand-bed ephemeral streams of central Australia, Sedimentology, v. 17, p. 1-40.

Williams, P.F. and Rust, B.R., 1969: The sedimentology of a braided river; J. Sediment. Petrol., v. 39, p. 649-679.

CHAPTER 5

MODERN ALLUVIAL FANS AND FAN DELTAS

Frank G. Ethridge
Department of Earth Resources
Colorado State University

Introduction

Alluvial fans constitute a distinct landform and depositional system characteristic of piedmont areas. Fisher and Brown (1972) describe alluvial fans as cone-shaped piles of sediment built where streams issue from a highland into an adjacent lowland. Fan deltas and clastic wedges are related features that belong under the same general heading of Fan Systems. Fan deltas are alluvial fans that build into a standing body of water (ocean, sea, lake, etc.). They generally have a distinct group of distal environments that owe their existance to this body of water. Proximal environments of fan deltas and alluvial fans are quite similar. Clastic wedges comprise a series of overlapping alluvial fan and/or fan delta deposits. The term clastic wedge has generally been applied to these thick sequences of coarse detrital units associated with fault block mountains.

Arid-Region Alluvial Fans

Bull (1962, 1963, 1964 and 1968) conducted extensive investigations of modern, arid and semi-arid region alluvial fans. These investigations are summarized in two later papers (Bull, 1972 and 1977). Morphologically arid-region alluvial fans are cone-shaped deposits that can be divided into three geomorphologic units: fan head (proximal fan), mid fan, and fan base (distal fan; Fig. 1C). Arid-region alluvial fans generally displays a concave longitudinal profile (Fig. 1C) and a

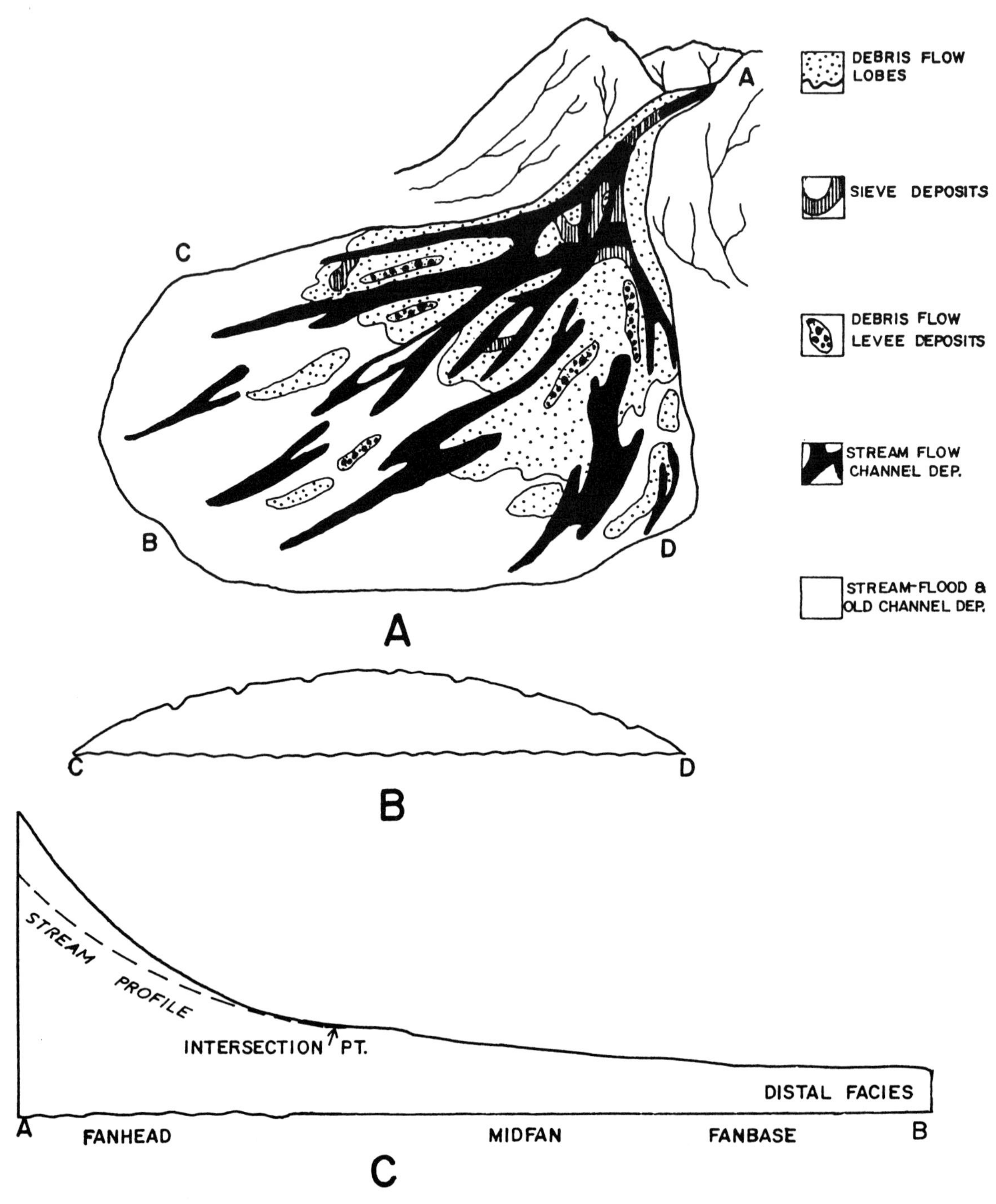

Figure 1. Distribution of sedimentary deposits (facies) and profiles of an idealized arid-region fan (modified from Spearing, 1975).

convex transverse profile (Fig. 1B). Fan surfaces slope from less than 1m to about 40m/km. Individual fans have lengths of 1 to 6 km and areas of 1 sq km to 900 sq km (Fisher and Brown, 1972; Spearing, 1975) although coalescing fans can form laterally extensive deposits parallel to a mountain front.

Sedimentary deposits of arid-region alluvial fans include sheetflood, stream-channel, debris-flow and sieve deposits (Fig. 1A). Temporarily entrenched stream channels on arid region fans are commonly filled with gravel that may be imbricated, massive or thick bedded. Deposits of stream channels are more poorly sorted than adjacent sheets of water-laid sediments deposited by very shallow braided distributaries. Low bars and very shallow (.1 to .5 m deep) braided distributary channels characterize the surface of the fan downfan from entrenched channels. Sheetflood deposits, resulting from deposition in this fan area consist of well sorted and massive to crossbedded and laminated silt, sand or gravel. Debris-flow deposits, composed of very poorly-sorted, unstratified, matrix-supported gravels (Fig. 2), are common on fans where the source area produces abundant fines (Fig. 3). The deposits lack clast imbrication and quicksand injection features may be present. In debris flows, the transport medium is mud, which moves more like a plastic than a liquid, so the flows have a high density and discosity and possess a finite strength for supporting larger clasts during transport. Debris flow levees may develop adjacent to channels in which some flows travel (Sharp, 1942; Fig. 3). Recognition of levee deposits in the rock record is difficult. Sieve deposits, a special type of fan stream deposit, lack sand, silt or clay. If fan surfaces are permeable enough to allow complete infiltration of water from a

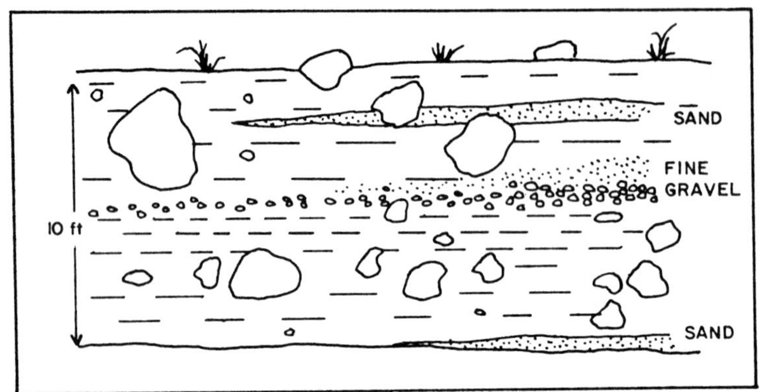

Figure 2. Internal characteristics of debris flow deposits near the apex of Miller Creek. Crude layering exists between successive flows (modified from Beaty, 1970).

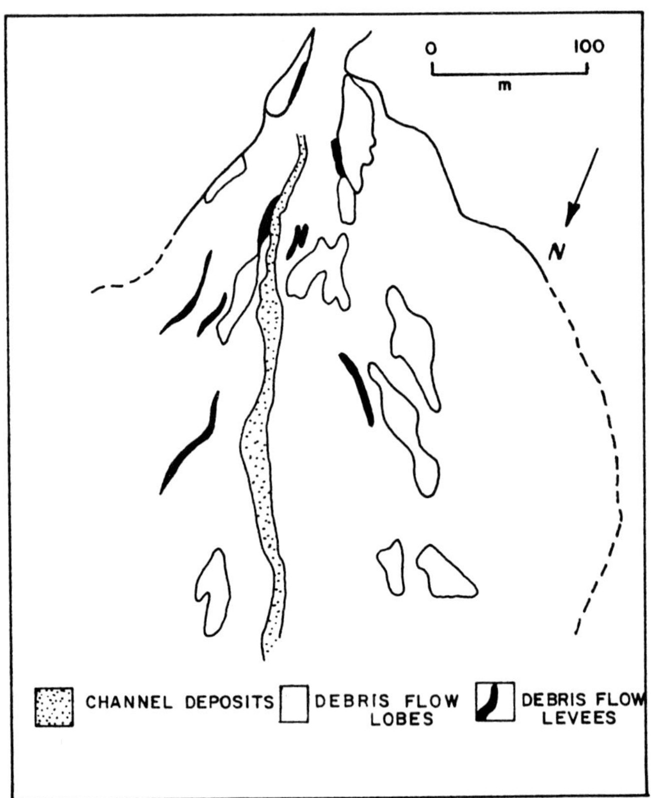

Figure 3. Diagramatic view of the upper Trollheim fan, Death Valley, showing the distribution of recent channel deposits, debris flow lobes and debris flow levees. There is a relief of about 100 meters over the area shown (modified from Hooke, 1967).

flood discharge, lobes of gravel, termed "sieve deposits" by Hooke (1967; Fig. 4), may form. In Death Valley fans sieve deposits are concentrated along the mid fan area at the intersection points of fan channels (Fig. 5). Few of these deposits have been recognized in the rock record, however.

Humid-Region Alluvial Fans

Humid-region alluvial fans are sub-divided into humid-glacial, humid-tropical, and humid-temperate (Kochel and Johnson, 1984). Modern humid-glacial fans on Alaska's southern coast have been studied by Reimnitz (1966), Boothroyd (1972), Boothroyd and Ashley (1975), Boothroyd and Nummedal (1978) and Galloway (1976). Some of these fans qualify as fan deltas and they will be discussed further in the next section. Humid-glacial fans range in areal extent from relatively small ones such as the Scott glacial outwash fan (24 km long by 2.5 to 11 km wide; Fig. 6) to the very large Copper River fan (50 km long by 25-100 km wide). Streams processes dominate, but discharge is seasonal, tied to meltwater runoff periods. Proximal to distal variations in braided bar morphology are described by Boothroyd (1972; Fig. 6). Sedimentary structures resulting from deposition of gravel sized-deposits on sheet bars in the upper fan area include crude horizontal bedding, clast imbrication, pebble strips (interpreted as relic antidune bedforms) and rare planar crossbeds (formed by flow separation along the edge of sheet bars). Down fan the sand content and bed relief increase and bar types change to in channel longitudinal bars. Sedimentary structures include horizontal bedding, large-scale cross beds and ripple cross lamination. The mid and lower fan areas are generally vegetated and active bradided streams are confined to narrow

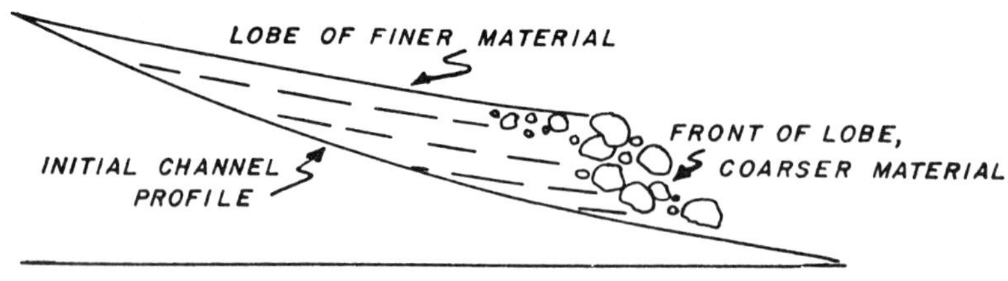

Figure 4. Schematic view of sieve lobe (modified from Hooke, 1967).

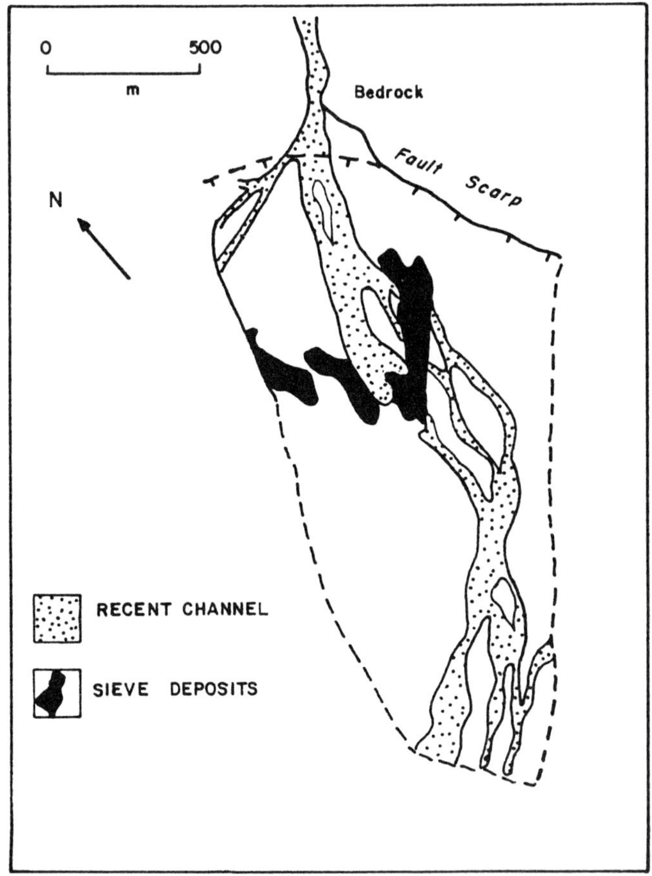

Figure 5. Diagrammetic view of the Gorak Shep fan, Death Valley, showing the concentration of sieve deposits at the intersection point of the fan channels in the mid fan area. There is a relief of about 250 meters from fan head to fan base (modified from Hooke, 1967).

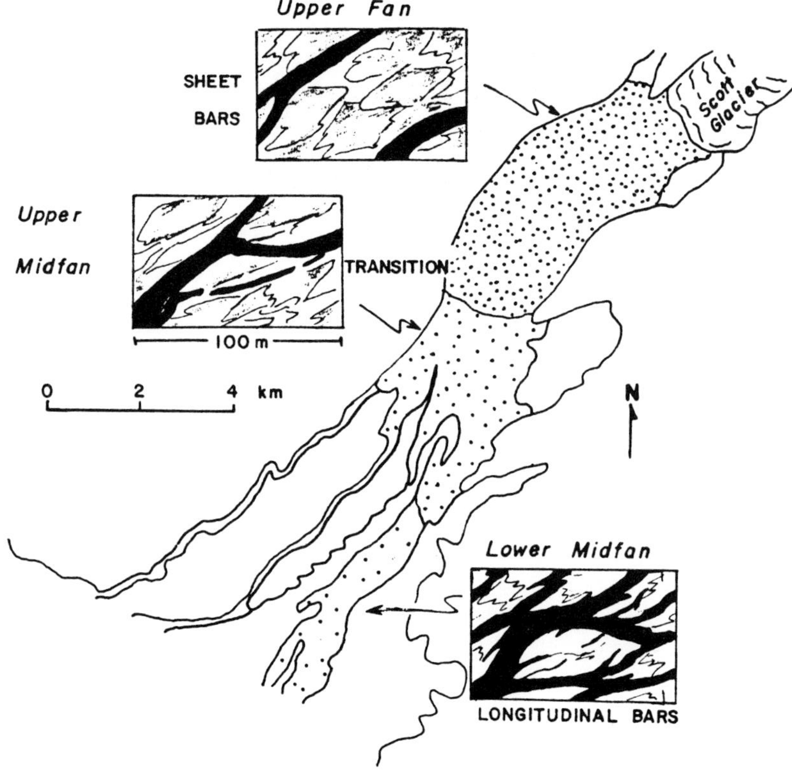

Figure 6. Diagrammetic view of the Scott outwash fan, southern Alaska, showing variations in bar morphology from fan head to fan base (modified from Boothroyd, 1972).

strips along the fan surface. Braided reaches on the lower fan reveal a combination of longitudinal and linguoid bars. Point and lateral bars develop where flow is confined between cohesive banks.

Descriptions of modern humid-tropical fans are limited to the Kosi fan in India (Gole and Chitale, 1966; Inglis, 1967; and Mukerji, 1976). Humid-tropical fans such as the Kosi extend for 160 km with widths that vary from 10 to 110 km (Fig. 7). Unfortunately, the details of sedimentary processes and deposits on humid-tropical region fans are unknown. Generally runoff is related to monsoonal rains and braided streams appear to be the major sediment transporter. Gradients are much

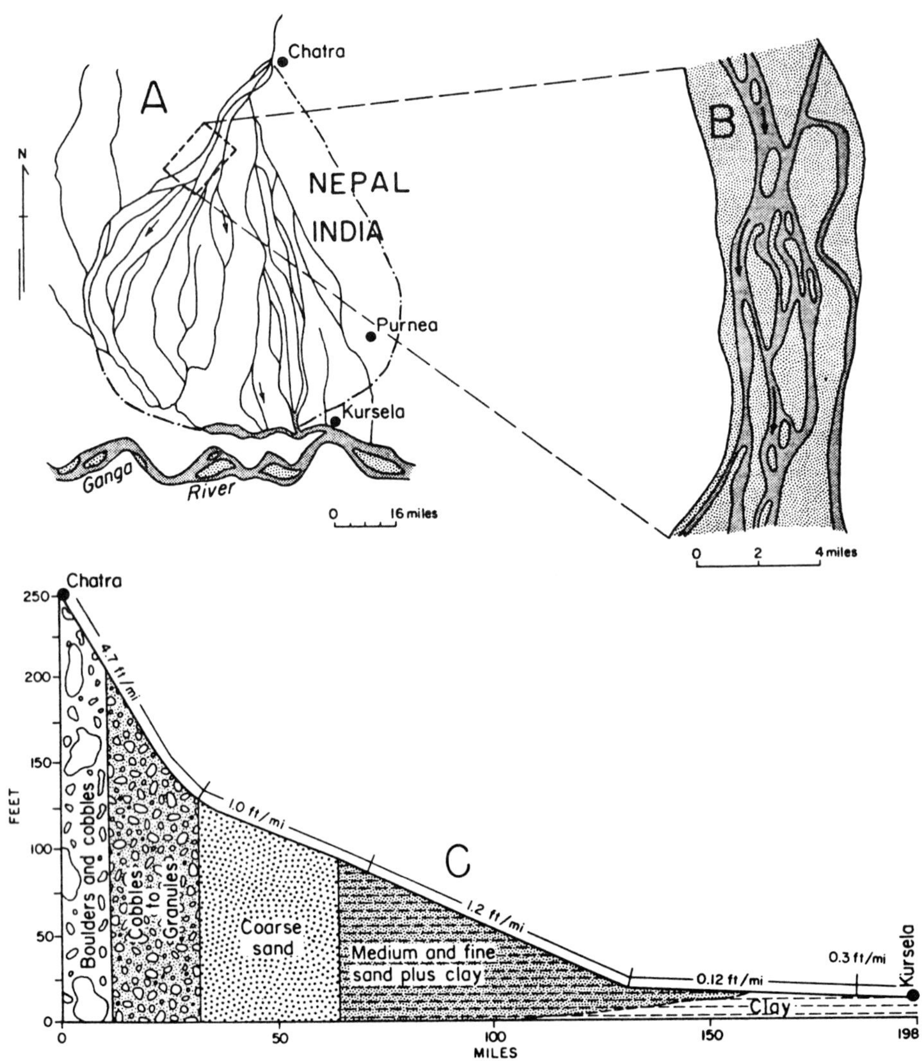

Figure 7. The Kosi River fan of India; a modern humid-tropical fan. (A) map view of fan. (b) braided stream pattern. (C) fan head to fan base profile showing decrease in slope and grain size (from McGowen, 1979; after Gole and Chitale, 1966).

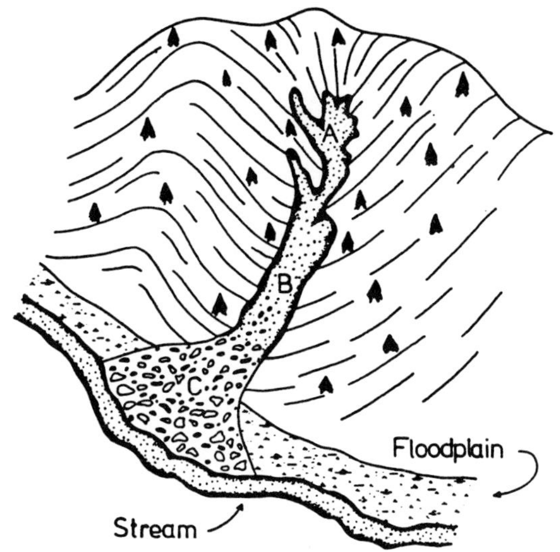

Figure 8. Diagrammatic illustration of a humid-temperate alluvial fan, Nelson county, Virginia. Area A is a bedrock scare of debris avalanche. Area B is an erosional chute. Area C is debris-dominated alluvial fan (modified from Kochel and Johnson, 1984).

lower (averaging less than 4.9m/km) than those found on arid-region fans and average grain size varies from boulders and cobbles at the fan head to clay at the fan base.

Modern humid-temperate fans in central Viriginia, described by Kochel and Johnson (1984), are small (typically less than 1 sq km) and thin (usually less than 20 m; Fig. 8). The major sedimentary process appears to be debris flows or debris avalanches which are triggered by intense rainfall. Similar debris flows have been described by Pierson (1980) on humid-temperate fans in New Zealand. Limited radiocarbon dating from the Viriginia fans suggests that the return interval on debris-avalanche events is 3000 to 6000 years and that 20 to 70% of the fan surface is activated during each event. Debris deposits, which consist of poorly-sorted, mud-supported gravels, show inverse grading and poorly-developed stratification. Vertical variations in texture and mineralogy are sharp and bed thicknesses (50 cm to 5 m) show no consistent down fan trend (Kochel and Johnson, 1984).

Recognition of Alluvial Fans

Alluvial fan deposits, generally characterized by coarse-grained, poorly-sorted, immature and locally derived gravels and sands, form lenticular sedimentary units with scoured channeled surfaces. Grain size and number and thickness of fine-grained units decrease and particle roundness increses downfan on most large alluvial fans. Fan geometries range from wedge-shaped to sheetlike. Current directions, as indicated by sedimentary structures display a radial flow pattern from the fan head area (Howard, 1966). Ideal vertical-sequence profiles through modern alluvial fan deposits are difficult to construct because of the variety of processes, climatic conditions and source rocks involved. The cyclic nature of vertical grain-size trends and sequences of sedimentation units in alluvial fan deposits reflect extrabasinal controls such as basin margin fault reactivation and autocyclic controls such as flood events (Heward, 1978; Gloppen and Steel, 1981; Nilsen, 1982; and Galloway and Hobday, 1983). Large-scale (hundreds to thousands of meters thick), fining-upward cycles (Fig. 9A) may reflect initial vigorous influx of coarse clastics after basin-margin fault reactivation followed by scarp retreat and lowering of relief during deposition. Large-scale, coarsening-upward cycles (Fig. 9B) reflect growth of the fan during major or persistent faulting/uplift. Rust and Koster (1984) suggest that the most common response to basin tectonis would be an upward-coarsening cycle representing fan progradation, followed by and upward-fining cycle representing a return to equillibrium. Gloppen and Steel (1981) describe an example of ancient deposits with these characteristics. Small-scale (a few to tens of

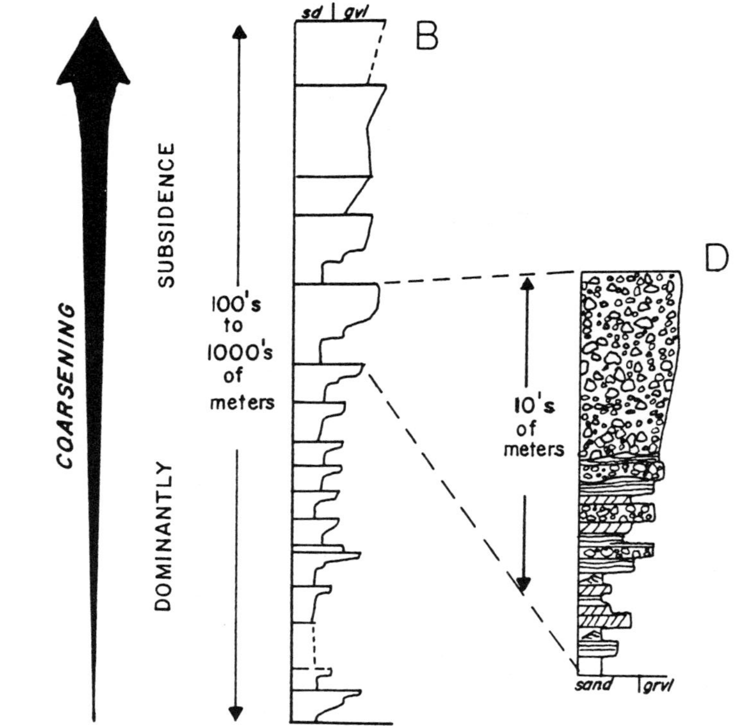

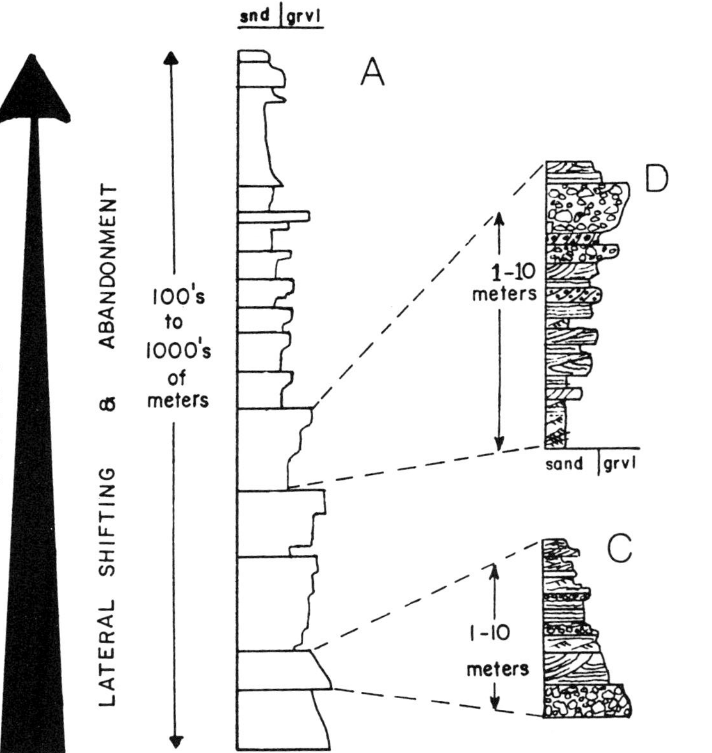

Figure 9. Idealized vertical sequences in alluvial fan deposits: (A) large-scale fining-upward sequence; (B) large-scale coarsening-upward sequence. (A') small-scale fining-upward sequence; (B') small-scale coarsening-upward sequence. See text for possible interpretations (based in part on data and discussions of Heward, 1978 and Gloppen and Steel, 1981).

meters thick), fining-upward cycles (Fig. 9C) with channelized bases probably reflect bar processes or braided channel abandonment and filling. Small-scale, coarsening-upward cycles (Fig. 9D) with non-channelized bases indicate progradation of individual fan lobes.

Fan Deltas

Fan deltas are alluvial fans that prograde into a standing body of water (Fig. 10). The essential elements, necessary for the development of fan deltas, are high relief adjacent to a coastal zone and high-gradient, bed-load streams that are usually braided to the coast. The resulting fan-shaped deposit of gravel and coarse sand-sized sediment contains distinctive subaerial, transitional and subaqueous components (Wescott and Ethridge, 1980 and Ethridge and Wescott, 1984). Fan morphology develops best along microtidal coastal zones which are characterized by abundant seasonal rainfall, although fan deltas are known to occur in macrotidal areas such as the Bristol Channel, Great Britain (Holmes, 1965) and Cook Inlet, Alaska (Hayes and Michel, 1982). Because of the unique circumstances necessary for their formation, fan deltas are usually confined to tectonically-active areas such as divergent and convergent plate margins, strike-slip margins and the margins of fault-bounded intracratonic seas and lakes.

Fan-delta deposits exhibit a distinct distal facies formed by the interplay of fluvial and marine-lacustrine processes, which distinguishes them from alluvial fans. In most fan-delta deposits that form at the edge of a marine basin this distinction involves the recognition and differentiation of beach gravels and fluvial gravels (Dobkins and Folk, 1970; Clifton, 1973; Wescott and Ethridge, 1980; Ethridge and Wescott, 1984; and Nemec and Steel, 1984). Table 1 details

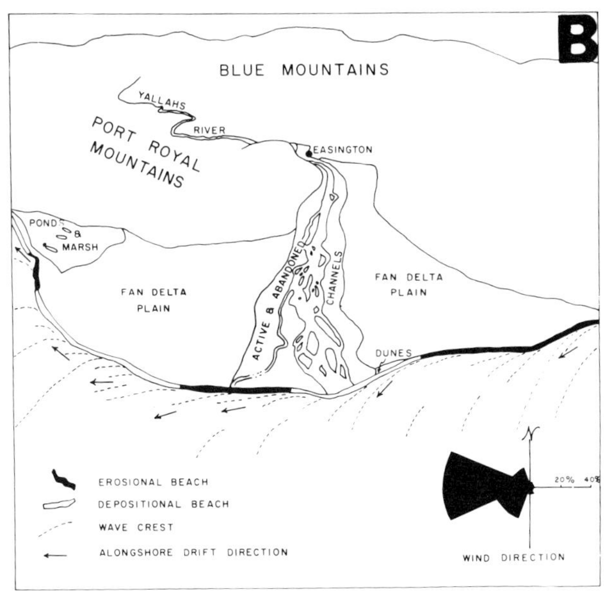

Figure 10. Yallahs fan delta southeastern Jamaica. (A) oblique aerial view looking north (photograph by J.S. Tyndale-Biscoe, Kingston, Jamaica). The Yallahs River is confined by artificial levees north of the coastal highway and ford across Yallahs River. Coastal highway runs east-west just north of real estate subdivisions on either side of the Yallahs River. (B) Interpretative sketch map of Yallahs fan delta (from Wescott and Ethridge, 1980).

Table 1. Criteria for distinghishing beach and fluvial channel gravels (based on part on data from Dobkins and Folk, 1970, and Clifton, 1973; after Ethridge and Wescott, 1981).

	BEACH GRAVELS	CHANNEL GRAVELS
* 1.	Well sorted	Poorly sorted
* 2.	Well segregated beds	Poorly segregated beds
3.	Continuous beds	Lenticular beds
* 4.	Gravels interbedded with gravelly sandstone - rare	Gravels interbedded with gravelly sandstone - common
* 5.	Erosional basal contacts - rare	Erosional basal contacts - common
* 6.	Repeated small-scale fining - up sequences - rare	Repeated small-scale fining - up sequences - common to rare
* 7.	Maximum clast size smaller than in adjacent channels	Maximum clast size larger than in adjacent beach
* 8.	Clayey or coaly laminae- rare	Clayey or coaly laminae- common to rare
9.	Imbrication - seaward dipping	Imbrication - landward dipping
10.	Sphericity - low Roundness - high	Sphericity - high Roundness - low
*11.	Horizontal beds of different size gravels or swash laminations	Horizontal beds or high angle through cross-beds

* = Criteria that can be identified in core

the criteria used for differeniating beach and fluvial gravels. In lacustrine and some marine fan-delta deposits the fan delta/alluvial fan distinction involves the recognition of gilbert-type gravel forest beds. Another important aspect in the recognition of fan-delta deposits is the differentiation of fan delta and submarine fan deposits. Criteria for differentiation of these two depositional systems is reviewed in Tables 2 and 3 (Howell and Link, 1979; Harbaugh and Dickinson, 1981; Harms and McMichael, 1983; Wescott and Ethridge, 1983; and Ethridge and Wescott, 1984).

Three depositional models, slope, shelf and Gilbert-type, are necessary to explain the various fan-delta deposits. A generalized slope model, based on studies of the Holocene Yallahs fan delta in southeastern Jamaica (Wescott, 1979; Wescott and Ethridge, 1980; and Ethridge and Wescott, 1984) is shown in Figure 11. The model characterizes fan deltas that prograde onto an island or continental slope. The overall coarsening-supward trend exhibited by most fan-delta deposits is not always obvious in slope fan deltas because of the presence of coarse-grained slump deposits associated with the heads of submarine canyons which are commonly developed in the lower portion of the sequence (Fig. 11). Slope-type fan deltas form along rift or divergent plate margins, pre-collision convergent plate margins and oceanic strike-slip margins (i.e., where continental or island shelves are narrow or absent). A generalized shelf model, based on numerous studies of fan-deltas along the southern Alaskan coast (Reimnitz, 1966; Boothroyd and Ashlety, 1975; Galloway, 1976; and Hayes and Ichel, 1982) is shown in Fig. 6, Chapter 2. Shelf-fan deltas often develop better coarsening-upward sequences than do slope-type fans. Numerous variants

Table 2. Comparison of submarine fan and fan-delta deposits (modified after Harbaugh and Dickinson, 1981).

Submarine fan	Fan-delta
1. Slope deposits - mainly hemipelagic lutites; rare lenticular, massive channel conglomerates and sandstones or olistostromal mudstones and sedimentary breccias	1. Subaerial delta plain deposits resemble alluvial fan deposits and usually consist of braided stream conglomerates and sandstones and/or coarse-grained meanderbelt deposits. Fine-grained floodplain deposits are rare. In arid regions debris flow and/or sieve deposits may constitute a significant fraction of the proximal deposits.
2. Inner fan deposits - one or a more discrete channel conglomerates and sandstones with erosional bases and fining and thinning-upward sequences grading into overbank turbidite sandstones and lutites; bulk of deposits consist of hemipelagic lutities interbedded with thin tubidite sandstones.	2. Delta front and shallow submarine fringe deposits - depending upon the interplay of fluvial and marine processes and climatic factors, deposits may consist of beaches, spits, tidal flats, shallow marine bars, delta foresets and/or carbonate reefs or algal mounds.
3. Mid-fan deposits - bulk of deposits consists of repeated thinning-upward cycles on turbidite sandstones resulting from deposition in multiple shallow braided channels.	3. Shelf and slope deposits - in shallow seas shelf deposits consist of muds interbedded with shallow nearshore submarine bar sandstones. Fans that build to the shelf edge overlie slope deposits similar to those described in (1) under submarine fans.
4. Outer fan deposits - characteristic deposits consists of hemipelagic lutites and repeated thickening-upward cycles of turbidite sandstone beds which have sheet-like geometries.	

Table 3. Major sedimentologic trends from nonmarine fan-data deposits to deep-water submarine fan deposits. Generalized from data and discussion by Howell and Link (1979), Harms and McMichael (1983) and Wescott and Ethridge (1983).

CRITERIA	NONMARINE FAN DELTA	SHALLOW MARINE	SUBMARINE FAN
1. Relative proportion of conglomerate to sandstone, mudstone, and shale	100% ————	————	———— <1%
2. Grading in conglomerates	Commonly ungraded	————	Commonly normal to inverse grading.
3. Texture of conglomerates	Clast supported (maybe matrix supported only in some arid region fans).		Matrix supported in slope and very proximal submarine feeder channels, otherwise clast supported.
4. Imbrication	Landward dipping imbrication (well developed).	Seaward dipping in beach conglomerates.	Landward dipping (some beds show no imbrication).
5. Orientation of long axes of clasts	Random, parallel and perpendicular.		Mostly parallel with some perpendicular.
6. Rip - up clasts			
a. size	Small		Variable
b. shape	Well rounded		Angular

Table 3. (continued)

CRITERIA	NONMARINE FAN DELTA	SHALLOW MARINE	SUBMARINE FAN
7. Sedimentary Structures	Mostly massive to crude horizontal bedding in coarse conglomerates; plannar or trough cross beds or horizontal beds in sandy conglomerates and conglomeratic sandstones.	Beach conglomerates and sandstones have horizontal beds and swash laminae.	Slump and fold strata in proximal slope and feeder channel deposits along with massive to graded conglomerates; Bouma sequences, flat laminations and rippled strata in top of preserved sequences in mid to distal fan sequences.
8. Fan slopes		Fan-delta fringe deposits composed of finer-grained material deposited on prodelta slopes. Depending upon the dominant transport mechanism, slope angles can range from a few degrees or less to very steep angle of respose slopes.	No mechanism for building and prograding steep slopes Modern submarine fans have very low average gradients; slopes of only a few minutes for larger fans to only slightly over one degree for smaller fans are common.
9. Other criteria	Evidence of subaerial exposure including; mudcracks, roots, eolian deposits, etc.; mudstones w/caliche horizons (a function of climate); common occurrence of plant fragments; absence of marine fossils in coarser grained units.	Marine fossils and abundant vertical and/or inclined burrow structures in sandstones interbedded with conglomerates.	Abundant deep water microfauna; abundant grazing traces on bedding plan exposures; transported fauna in coarser grained units.

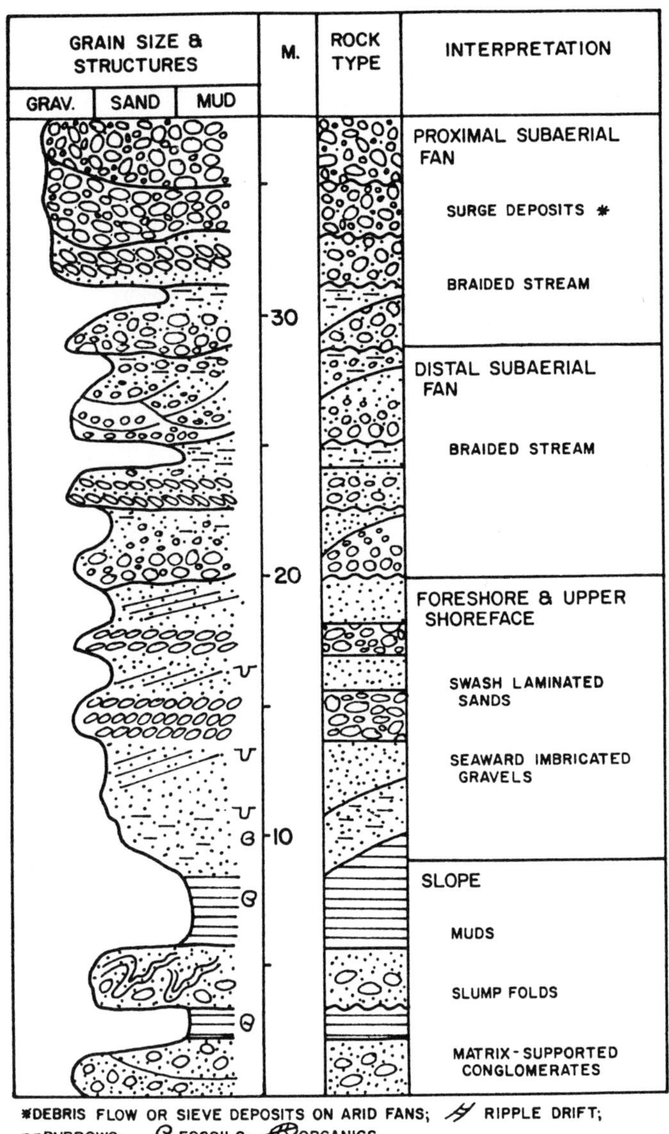

Figure 11. Vertical sequence in a hypothetical slope-type fan delta based on studies of the Yallahs fan delta and slope, Jamaica (from Ethridge and Wescott, 1984).

on the shelf model presented here exist. Holocene and ancient fan deltas along divergent plate margins and intracratonic rifts often show coarse-grained terrigenous clastic fan deposits associated with carbonate reefs or algal mounds. Brown (1979) presents an excellent model for this type of fan-delta system. Gilbert-type fan deltas, as described in the Pleistocene Lake Bonneville (Gilbert, 1890), have also been recognized in ancient, fault bounded, intracratonic basins in Colorado and New Mexico (Casey and Scott, 1979; Walker and Harms, 1980; and Millberry, 1983). The deltas are characterized by the presence of large-scale, high-angle, gravel foreset beds (Fig. 12), which probably relate to delta progradation into a shallow-water interdeltaic embayment. Saline waters of the embayment were probably displaced during a flood resulting in homopycnal flow locally (Casey and Scott, 1979).

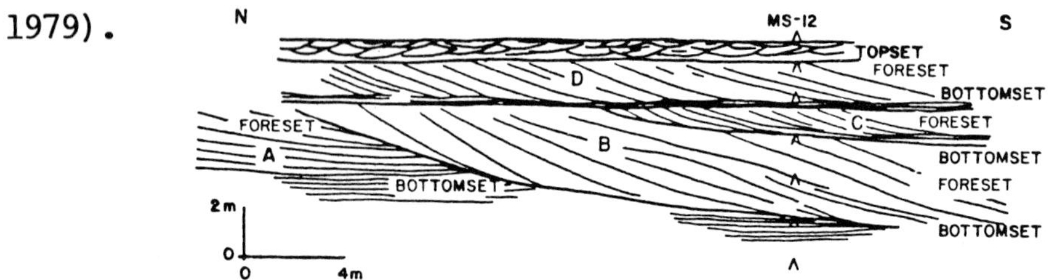

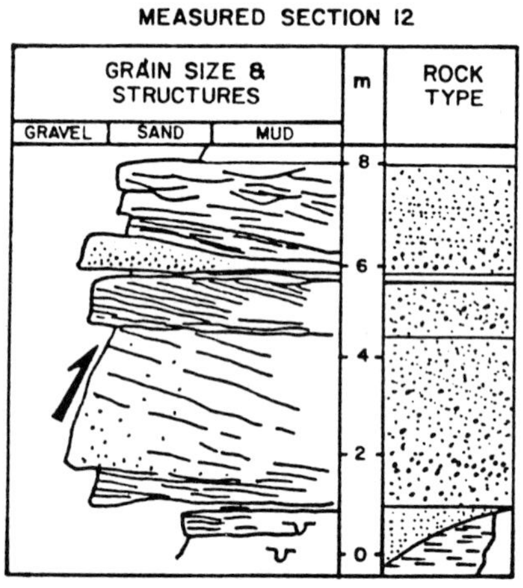

Figure 12. Outcrop sketch and measures section of topset, foreset and bottomset beds of Gilbert-type, marine fan-delta deposits, Pennsylvanian, Taos Trough, New Mexico (after Casey and Scott, 1979, Fig. 5).

REFERENCES

Beaty, C.B., 1970, Age and estimated rate of accumulation of an alluvial fan, White Mountains, California, USA: Amer. Jour. Sci., v. 268, p. 50-77.

Boothroyd, J.C., 1972, Coarse-grained sedimentation on a braided outwash fan, northeastern Gulf of Alaska: Columbia, S.C., Univ. of South Carolina, Coastal Res. Div., Tech. Rept. No. 6-CRD, 127 p.

Boothroyd, J.D. and Ashley, G.M., 1975, Processes, bar morphology and sedimentary structures on braided outwash fans, northeastern Gulf of Alaska, in, Mcdonald, B.C., and Jopling, A.V. (eds.), Glacio-fluvial and glaciolacustrine sedimentation: Soc. Econ. Paleo. and Mineral. Sp. Pub. 23, p. 193-222.

Boothroyd, J.C. and Nummedal, D., 1978, proglacial braided outwash: a model for humid alluvial fan deposits, in, Miall, A.D. (ed.), Fluvial sedimentology: Can. Soc. Petrol. Geol. Mem. 5, p. 641-668.

Brown, L.F., Jr., 1979, Deltaic sandstone facies of the Mid-Continent, in, Hyne, N.J., Pennsylvanian sandstones of the mid-continent: Tulsa Geol. Soc. Sp. Pub. No. 1, p. 35-63.

Bull, W.B., 1962, Relation of texture (CM) patterns to depositional environments of alluvial fan deposits: Jour. Sed. Petrology, v. 32, p. 211-216.

Bull, W.B., 1963, Alluvial-fan deposits in western Fresno County, California: Jour. Geol., v. 71, p. 243-250.

Bull, W.B., 1964, Geomorphology of segmented alluvial fans in western Fresno County, California: U.S. Geol. Survey Prof. Paper 252, p. 89-129.

Bull, W.B., 1968, Alluvial fans: Jour. Geol. Educ., v. 16, p. 101-106.

Bull, W.B., 1972, Recognition of alluvial-fan deposits in the stratigraphic record, _in_, Rigby, J.K. and hamblin, W.K. (eds.), Recognition of ancient sedimentary environments: Soc. Econ. Paleon. and Mineral., Spec. Pub. No. 16, p. 63-83.

Bull, W.B., 1977, The alluvial fan environment: Progress in Physical Geography, v. 1, p. 222-270.

Casey, J.M. and Scott, A.J., 1979, Pennsylvanian coarse-grained fan deltas associated with the Uncompahgre uplift, Talpa, New Mexico: New Mexico Geol. Soc. Guidebook, 30th Field Conf., Santa Fe County, p. 211-218.

Clifton, H.E., 1973, Pebble segregation and bed lenticularity in wave-worked versus alluvial gravel: Sedimentology, v. 20, p. 173-187.

Dobkins, J.E., Jr. and Folk, R.L., 1970, Shape development on Tahiti Nui: Jour. Sed. Petrology, v. 40, p. 1167-1203.

Ethridge, F.G. and Wescott, W.A., 1984, Tectonic setting, recognition and hydrocarbon reservoir potential of fan-delta deposits, _in_, Koster, E.H. and Steel, R.J. (eds.), Sedimentology of gravels and conglomerates: Can. Soc. Petrol. Geol. Mem. 10, p. 217-235.

Fisher, W.L. and Brown, L.F., Jr., 1972, Clastic depositional systems - a genetic approach to facies analysis: annotated outline and bibliography: The Univ. of Texas at Austin Bur. of Econ. Geol., 211 p.

Galloway, W.E., 1976, Sediments and stratigraphic framework of the Copper River fan delta, Alaska: Jour. Sed. Petrology, v. 46, p. 726-737.

Galloway, W.E. and Hobday, D.K., 1983, Terrigenous clastic depositional systems: applications to petroleum, coal, and uranium exploration: Springer-Verlag, New York, 423 p.

Gilbert, G.K., 1890, Lake Bonneville: U.S. Geological Survey Monograph, v. 1, 438 p.

Gloppen, T.G. and Steel, R.J., 1981, The deposits, internal structure and geometry of six alluvial fan-fan delta bodies (Devonian, Norway) - a study in the significance of bedding sequences in conglomerates, in, Ethridge, F.g. and Flores, R.M. (eds.), Ancient nonmarine depositional environments: models for exploration: Soc. Econ. Paleo. and Mineral. Sp. Pub. 31, p. 49-69.

Gole, C.v. and Chitale, S.V., 1966, Inland delta building activity of the Kosi River: Proc. Amer. Soc. Civil Engrs., Jour. Hydraulics Div., HY2, v. 92, p. 111-122.

Harbaugh, D.W. and Dickinson, W.R., 1981, Depositional facies of Mississippian clastics, Antler Foreland Basin, central Diamond Mountains, Nevada: Jour. Sed. Petrology, v, 51, p. 1223-1234.

Harvey, A.M., 1984, Debris flows and fluvial deposits in Spanish Quaternary alluvial fans: implications for fan morphology, in, Koster, E.H. Steel, J.R. (eds.), Sedimentology of gravels and conglomerates: Can. Soc. Petrol. Geol. Memoir 10, p. 123-132.

Hayes, M.O. and Michel, J., 1982, Shoreline sedimentation within a forarc embayment, lower Cook Inlet, Alaska: Jour. Sed. Petrology, v. 52, p. 251-263.

Heward, A.P., 1978, alluvial fan sequence and megasequence models, with examples from Westphalian D-Stephanian B coalfields, northern Spain, in, Miall, A.D. (ed.), fluvial sedimentology: Can Soc. Petrol. Geol. Memoir 5, p. 669-702.

Holmes, a., 1965, Principles of Physical Geology, 2nd Edition: The Ronald Press Co., New York, 1288 p.

Hooke, R. LeB., 1967, Processes on arid-region alluvial fans: Jour. Geol., v. 75, p. 438-460.

Howard, J.D., 1966, Patterns of sediment dispersal in the Fountain Formation of Colorado: Mtn. Geologist, v. 3, p. 147-153.

Howell, D.G. and Link, M.H., 1979, Eocene conglomerate sedimentology and basin analysis, San Diego and the southern California borderland: Jour. Sed. Petrology, v. 49, p. 517-540.

Inglis, C.D., 1967, Inland delta building activity of Kosi River: Proc. Amer. Soc. Civil Engrs., Jour. Hydraulic Div., HY1, v. 93, p. 93-100.

Kochel, R.C. and Johnson, R.A., 1984, Geomorphology and sedimentology of humid-temperate alluvial fans, central Viriginia, in, Koster, E.H. and Steel, R.J. (eds.), Sedimentology of gravels and conglomerates: Can. Soc. Petrol. Geol. Mem. 10, p. 109-122.

Leckie, D.A. and Walker, R.G., 1982, Storm- and tide-dominated shorelines in Cretaceous Moosebar-Lower Gates interval-outcrop equivalents of Deep Basin gas trap in western Canada: Amer. Assoc. Petrol. Geol. Bull., v. 66, p. 138-157.

McGowen, J.H., 1979, Alluvial fan systems, in, Galloway, W.E., Kreitler, C.W. and McGowen, J.H. (eds.), Depositional and ground-water flow systems in the exploration for uranium: The Univ. of Texas Bur. of Econ. Geol., p. 43-79.

McGowen, J.H. and Bloch, S., 1985, Depositional facies, diagnesis and reservoir quality of Ivishak Sandstone (Sadlerochit Group), Prudhoe Bay field (abs.): Amer. Assoc. Petrol. Geol. Bull., v. 69, p. 286.

Melvin, J. and Knight, A.S., 1984, Lithofacies, diagenesis and porosity of the Ivishak Formation, Prudhoe Bay area, Alaska, in, Mcdonald, D.A. and Surdam, R.C. (eds.), Clastic diagnesis: Amer. Assoc. Petrol. Geol. Memoir 37, p. 347-365.

Millberry, K.W., 1983, Tectonic control of Pennsylvanian fan-delta deposition, southwestern Colorado (abs.): Amer. Assoc. Petrol. Geol. bull., v. 67, p. 514.

Makerji, A.B., 1976, Terminal fans of inland streams in Sutlej-Yamuna plain, India: Zeitschrift fur Geomorphologie, v. 20, p. 190-204.

Nemec, W. and Steel, R.J., 1984, Alluvial and coastal conglomerates: their significant features and some comments on gravelly mass-flow deposits, in, Koster, E.H. and Steel, R.J. (eds.), Sedimentology of gravels and conglomerates: Can Soc. Petrol. Geol. Memoir 10, p. 1-31.

Nilsen, T.H., 1982, Alluvial fan deposits, in, Scholle, P.A. and Spearing, Darwin (eds.), Sandstone depositional environmetns, Amer. Assoc. Petrol. Geol. Mem. 31, p. 49-86.

Ojeda, H.A.O., 1982, Structural framework, stratigraphy and evolution of Brazilian marginal basin: Amer. Assoc. Petrol. Geol. Bull., v. 66, p. 732-749.

Pierson, T.C., 1980, Erosion and deposition by debris flows at Mt. Thomas, New Zealand: Earth Surface Processes, v. 5, p. 1952-1984.

Reimnitz, E., 1966, Late Quaternary history and sedimentation of the Copper River delta and vicinity, Alaska, Ph.D. Thesis, Univ. of California, San Diego, 160 p.

Rust, B.R. and Koster, E.H., 1984, Coarse alluvial deposits, in, Walker, R.G. (ed.), Facies Models (Sec. Edition), Geoscience Canada Reprint Series 1, p. 53-69.

Spearing, D.R., 1975, Summary sheets of sedimentary deposits: Geol. Soc. Amer., Pub. MC-8.

Sharp, R.P., 1942, Mudflow levees: Jour. Geomorphology, v. 5, p. 222-227.

Walker, T.R. and Harms, J.C., 1980, Fan-delta deposition, Minturn Formation, McCoy area, colorado: Unpublished field notes, Rky. Mtn. Sect. Soc. Econ. Paleo and Mineral. Fall Field Conf.

Wescott, W.A., 1979, The Yallahs fan delta, southeastern Jamaica: a depositional model for active tectonic coastlines: Ph.D. Thesis, Colorado State Univ., 178 p.

Wescott, W.A. and Ethridge, F.G., 1980, Sedimentology and tectonic setting of fan deltas - Yallahs fan delta, southeast Jamaica: Amer. Assoc. Petrol. Geol. Bull., v 64, p. 374-399.

CHAPTER 6

ANCIENT ALLUVIAL FANS AND FAN DELTAS

William E. Galloway

INTRODUCTION

Although much has been learned about modern alluvial-fan and fluvial systems, the step from active systems to their ancient geologic counterparts is a big one. The following notes attempt, by means of selected examples, to illustrate the facies characteristics of a broad spectrum of fluvial and fan deposits.

Alluvial fans are primarily geomorphic features most readily distinguished by their conical, lobate, or arcuate morphology. As morphology is not readily reconstructed in the stratigraphic record, indirect criteria are necessary. In addition to morphology, alluvial fans display several features suggestive of their origin (Fig. G-1):

1. Radial or arcuate sediment dispersal pattern reflecting the point source. Note sand distribution on Kern fan (Fig. G-1).
2. Highly compressed down-flow textural and compositional gradients. The Kern fan grades from dominantly gravel to less than 50 percent sand along a transverse section.
3. A uniquely predictable ground-water flow system that may produce distinctive geochemical facies or alteration patterns.
4. Well-defined areal relationships with structural features or known uplands.
5. Distinctive associations with bounding facies, such as lacustrine deposits or major through-flowing trunk stream sequences.

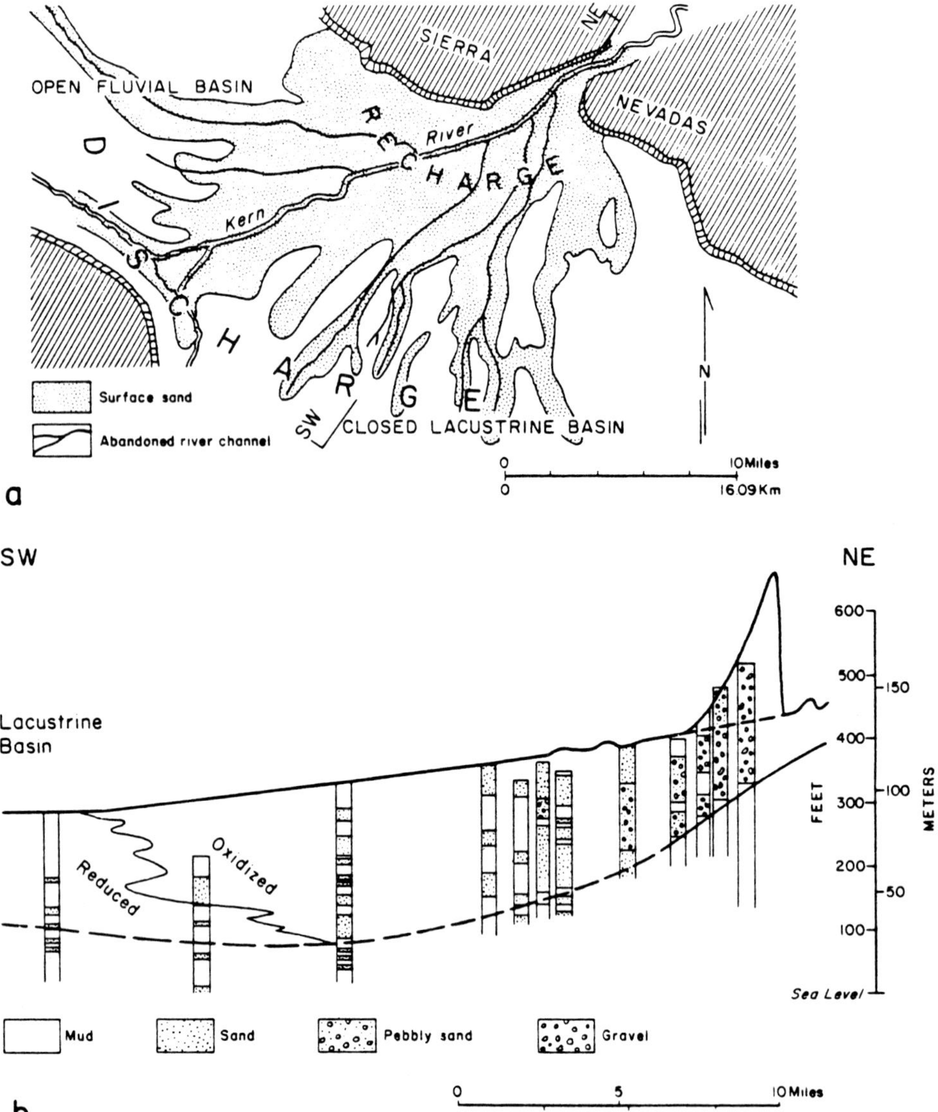

Figure G-1. Sedimentary and hydrologic features of the Kern wet alluvial fan, southern Great Valley, California: (A) map view showing present and former channel courses radiating from the Sierra Nevada Mountain front, (B) cross section through the fan illustrating the down-fan decrease in gravel and sand within the young fan sequence and the generalized position of oxidized and reduced fan deposits. (From Galloway, 1980).

Fan deltas are special situations in which alluvial fans build directly into a standing body of water. Although it might be argued that only small features qualify as true fan deltas, the juxtaposition of marine deposits with coarse fan-like sediments is worthy of note, and the term "fan delta" has been widely applied (see Ethridge, this volume).

Atokan Fan Deltas of Taos Trough

Pennsylvanian strata now exposed in the Sangre de Cristo Mountains record deposition of a major clastic wedge shed from the eastern margin of the ancestral Uncompahgre Uplift into the rapidly subsiding Taos Trough (Casey and Scott, 1979; Casey, 1980). These wedges record an Atokan to early Desmoinesian episode of uplift that resulted in the eastward and southeastward progradation of fan-delta systems into the basin (Figs. G-2, 3a, 3b, and 3c). Rapid lateral shifting of these small fans and fan deltas during basin subsidence produced a vertical sequence of complexly interbedded coarse clastics alternating with marine limestone and mudstone. Similar facies assemblages characterize many cratonic basins associated with development of the ancestral Rocky Mountains.

By Atokan time, vertical uplift along the Pecos-Picuris fault produced coarse alluvial-fan, braided-stream and fan-delta complexes along the western margin of the trough. Alluvial-fan and associated braided-channel deposits are characterized by ribbon to sheet sand and conglomerate unit geometries. Trough and tabular cross-stratification dominate internal structures (Fig. G-4a). These coarse units grade basinward into fan-delta sequences consisting of upward-coarsening

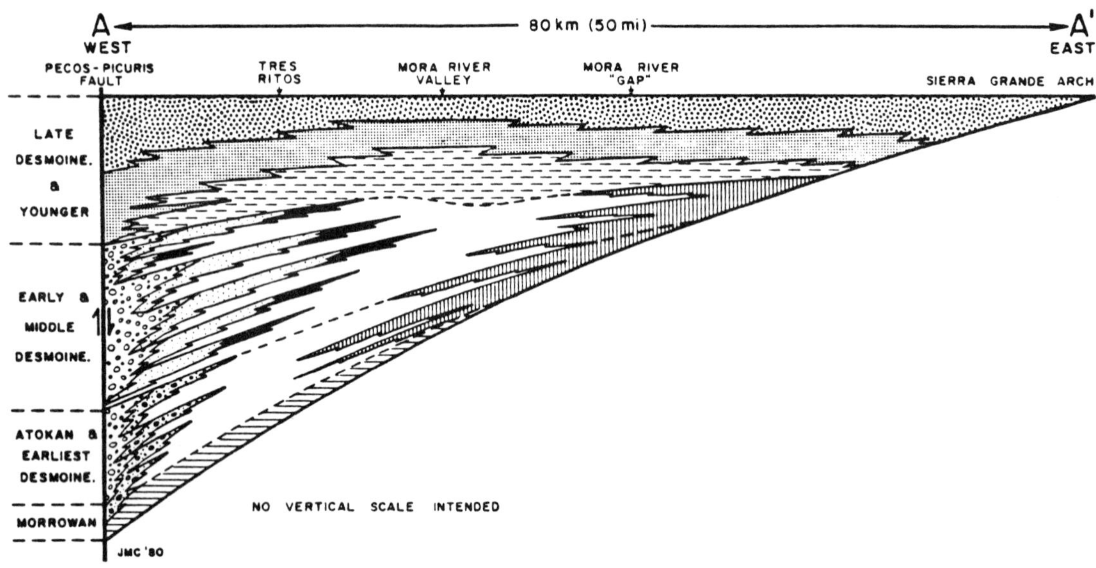

Figure G-2. Schematic cross section of the Pennsylvanian fill of the Taos Trough. (From Casey, 1980).

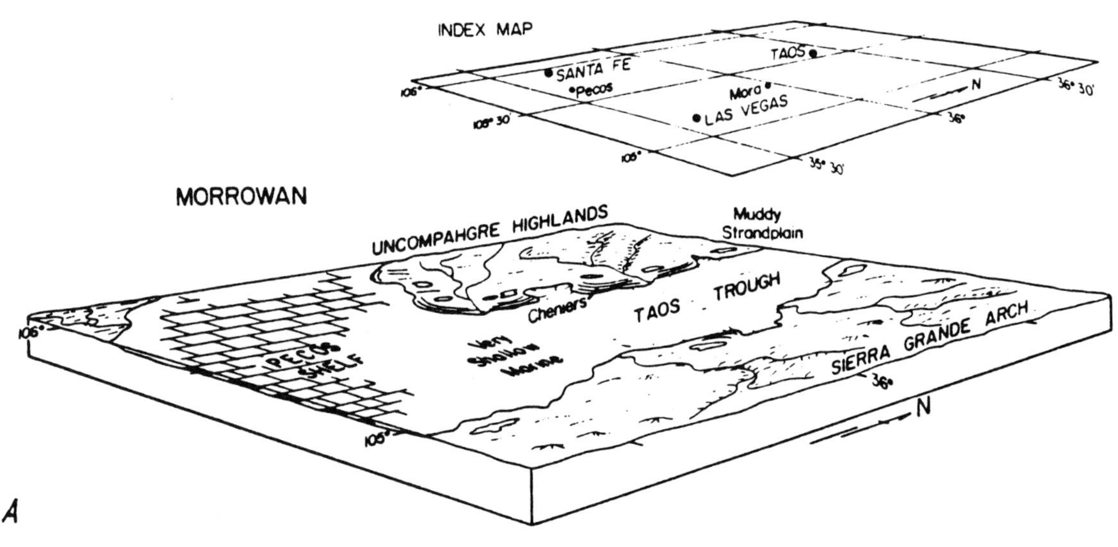

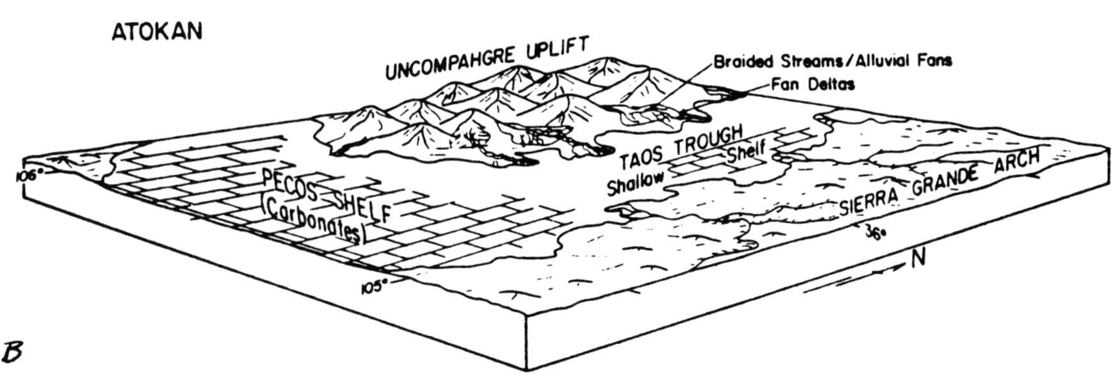

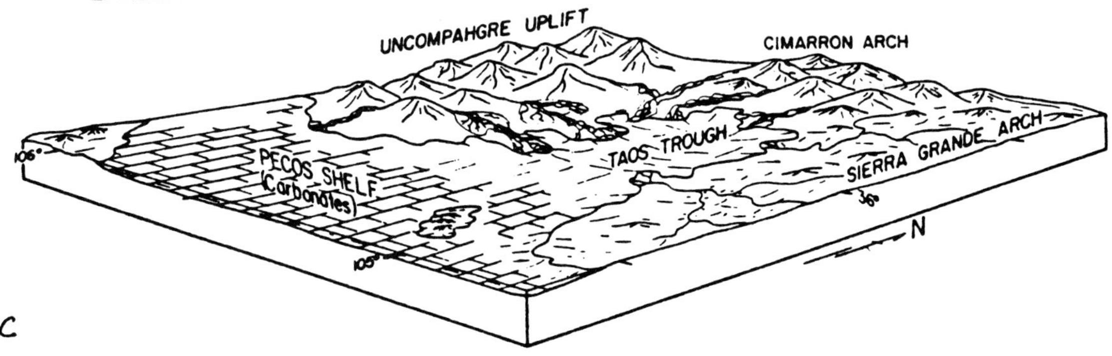

Figure G-3. Morrowan, Atokan, and earliest Desmoinesian paleogeography of the Taos Trough and surrounding areas. (From Casey, 1980).

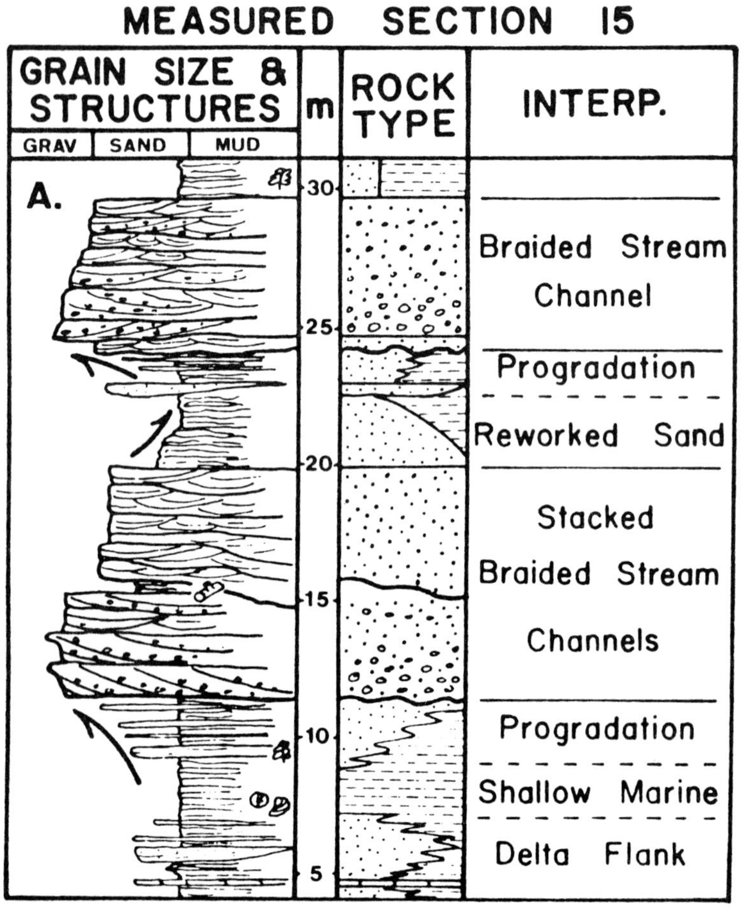

MEASURED SECTION 15

GRAIN SIZE & STRUCTURES
GRAV SAND MUD
m
ROCK TYPE
INTERP.

A.

30
25
20
15
10
5

Braided Stream Channel
Progradation
Reworked Sand
Stacked Braided Stream Channels
Progradation
Shallow Marine
Delta Flank

Figure G-4a. Fan-delta variations. (From Casey, 1980).

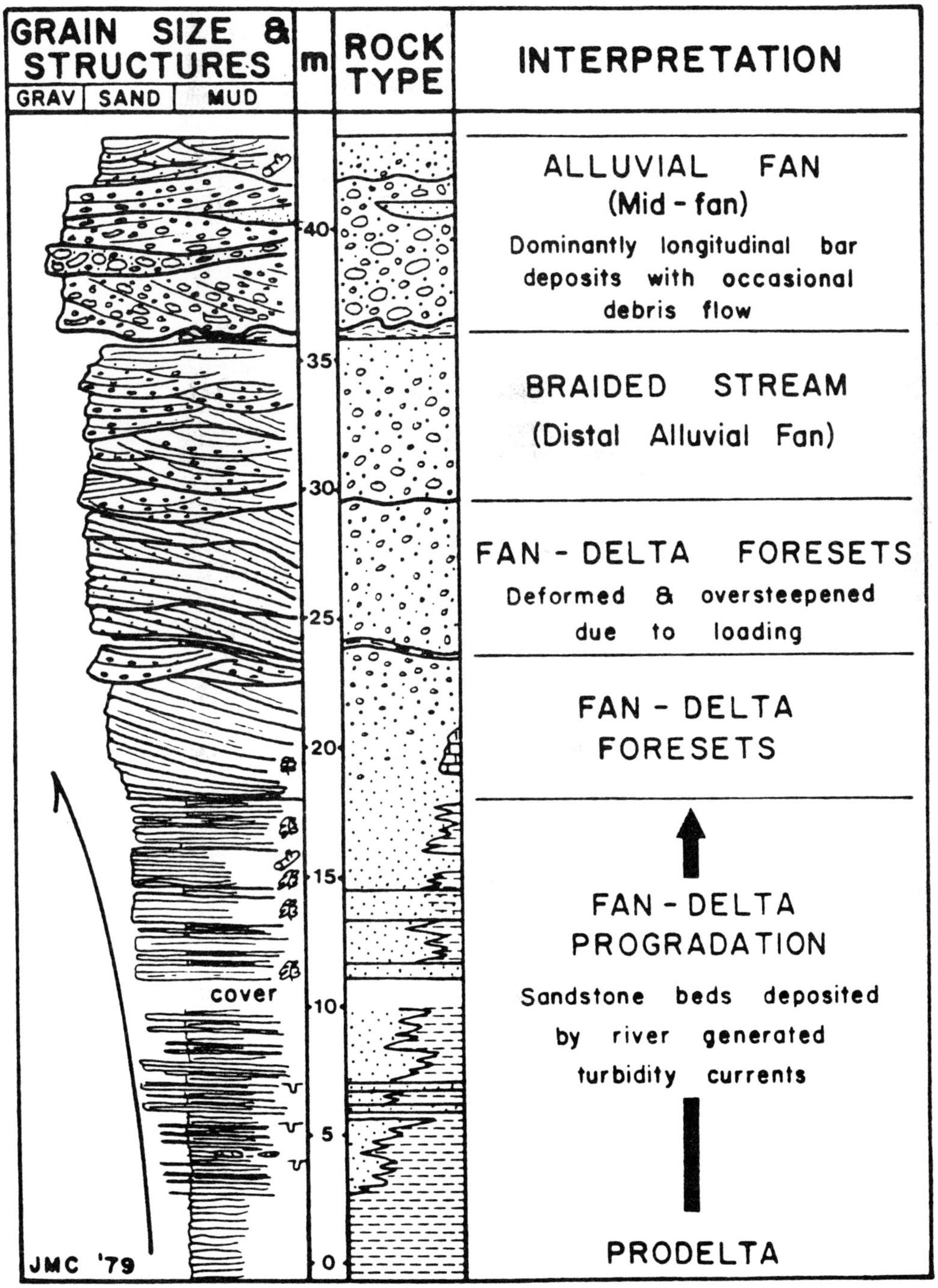

Figure G-4b. Preservation of complete fan-delta/braided-stream/alluvial-fan facies tract in vertical succession. (From Casey, 1980).

cycles (Fig. G-4b). Basinward fan deltas in turn grade into thick sequences of dark-gray, calcareous siltstone.

Ogallala Alluvial Apron

The Ogallala Formation is a Neogene alluvial apron that occurs east of the Rocky Mountains from South Dakota to the Southern High Plains of Texas. Because of its areal extent, scale, and more nebulous association with a tectonically active upland, the Ogallala illustrates problems in interpretation of alluvial fans in the stratigraphic record (Seni, 1980).

The Ogallala was deposited by coalescent, low-gradient, wet alluvial fans that display complex histories of erosion and deposition. Geometry and depositional facies of the Ogallala in north Texas have been determined by outcrop studies and synthesis of water-well driller's logs. An averaging technique was employed to compensate for the variability of driller's descriptions. Data from all logs in a 2.5-minute quadrangle area were compressed into a single average data point used for mapping.

Ogallala depositional systems and component sand-body geometry were delineated by construction of interval and net sand and gravel isopach maps (Fig. G-6) and sand and gravel percentage maps. The three-dimensional trends in thickness and coarse clastics distribution is consistent with a wet alluvial-fan model derived from large modern fans such as the Kosi (Table G-1).

Three overlapping fan lobes (Fig. G-5) were constructed by sediment transported along pre-Ogallala valleys that are as much as 200 ft (60 m) deep. Each lobe displays a general downdip decrease in thick-

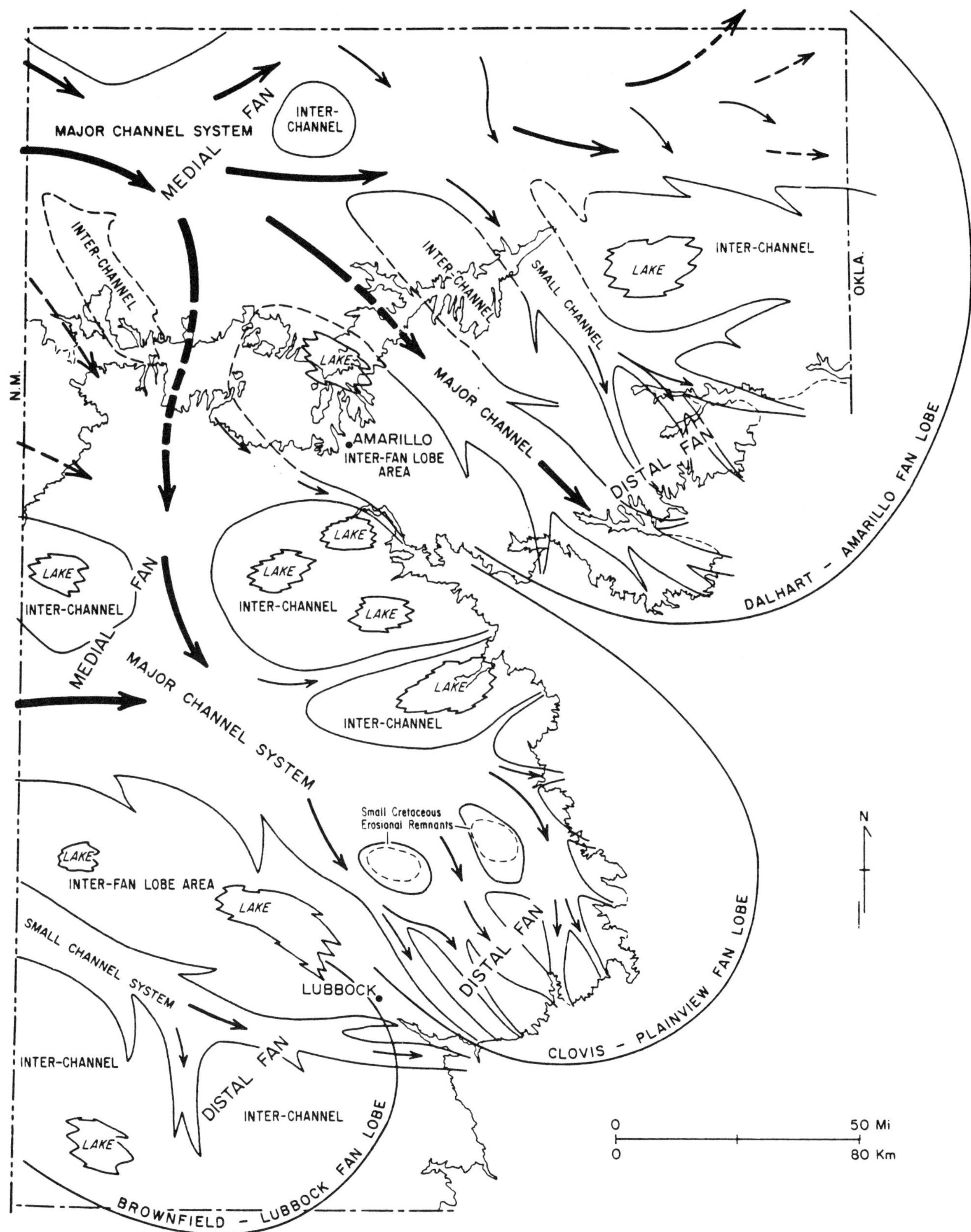

Figure G-5. Schematic illustration of Ogallala depositional facies and sediment dispersal systems. Width and length of arrows indicate relative intensity of fluvial processes. Lobes become younger to south. (From Seni, 1980).

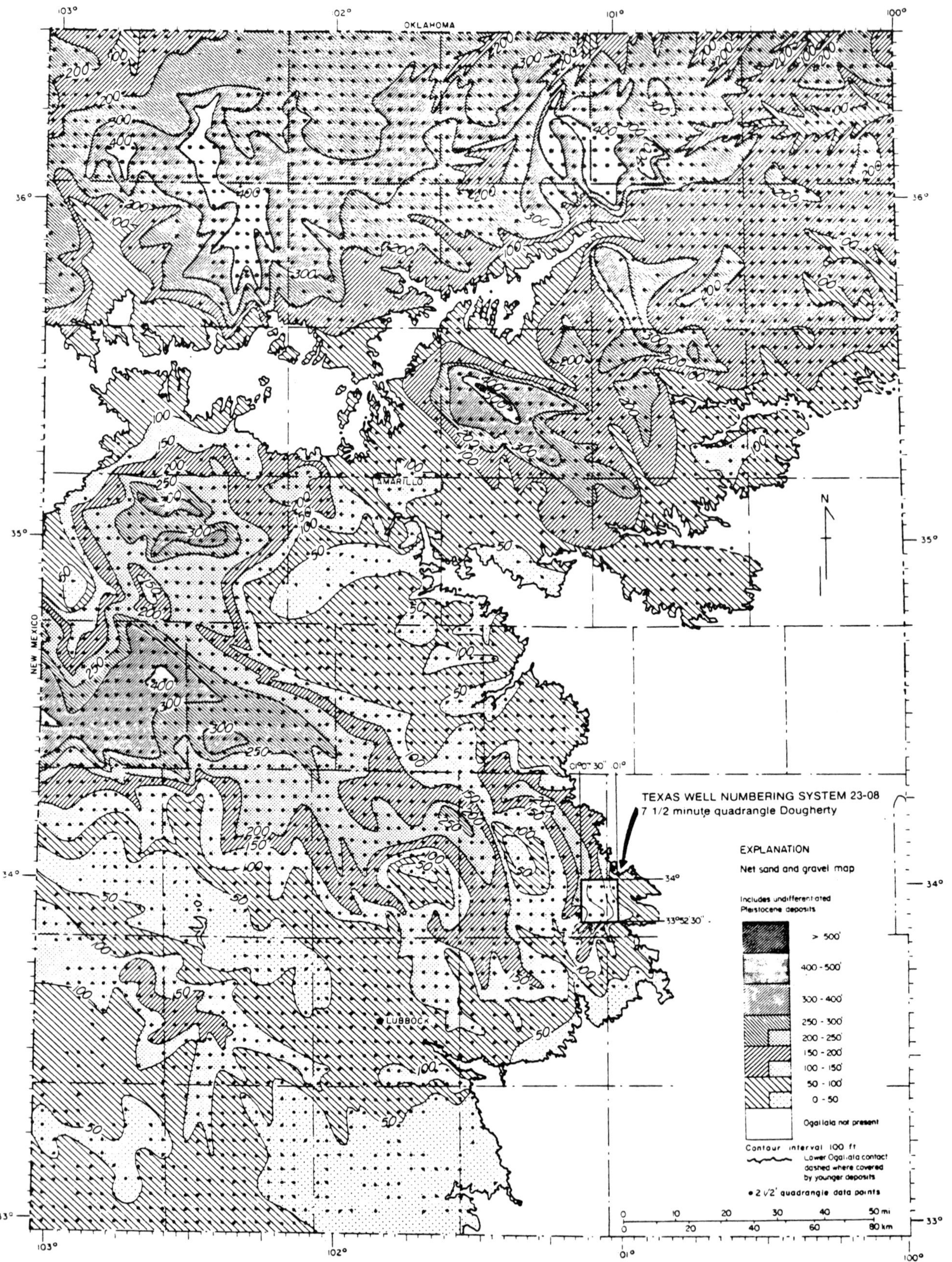

Figure G-6. Net sand and gravel isolith map, Ogallala Formation. Data points represent mean net sand and gravel of all wells in each quadrangle. Thin, undifferentiated Pleistocene deposits may be included. (From Seni, 1980).

Table G-1. Principal characteristics of Ogallala facies. (Seni, 1980).

Geomorphic unit	Environment / Processes	Textural trends (ft, m; grav, sand, mud)	Thickness and percentage sand and gravel	Sedimentary structures	Sand-body geometry
Medial alluvial fan	Wet (seasonal processes), low-gradient medial alluvial fan platform; braided stream; environment—contributary and trunk drainage	400 120, 300 90, 200 60, 100 30	Total thickness 200-500 ft 60-150 m Individual sheets 50-300 ft 15- 90 m Sand-gravel percentage 40-90%	Structureless—massive, imbricated pebbles, trough-fill cross-stratified, parallel laminae	Overall facies geometry—lobate, sheet-like, some valley-fill; 25-100 miles (40-160 km) wide; internal geometry—shallow, broad channel-fill lenses, coalesced into a sand-gravel sheet
Distal alluvial fan	Wet (seasonal processes), low-gradient distal alluvial fan; braided stream, distributary network	200 60, 100 30	Total thickness 100-400 ft 30-120 m Individual channels 25-200 ft 10- 60 m Sand-gravel percentage 40-90%	Structureless—trough-fill and foreset cross-stratified, parallel laminae, imbricated pebbles	Overall facies geometry—digitate, bifurcating downdip; linear belts 10-30 miles (15-45 km) wide; internal geometry—coalesced shallow, broad channel-fill lenses; channel systems thin and become finer grained downdip
Interchannel	Interchannel floodbasin, mud flat, lacustrine basin, eolian	200 60, 100 30	Total thickness 10-100 ft 3- 30 m Individual units 10-30 ft 3-10 m Sand-gravel percentage <20-40%	Structureless, mottled, burrowed, convoluted, parallel laminae; burrowed and parallel laminated limestone	Overall facies geometry—thin belts 5-20 miles (8-30 km) wide of low net and percentage sand-gravel in between channel systems in distal alluvial fan; interchannel areas broaden downdip as net and percentage sand-gravel decreases
Interfan	Interfan floodbasin, mud flat, lacustrine basin, eolian	200 60, 100 30	Total thickness 10-100 ft 3- 30 m Individual units 10-50 ft 3-15 m Sand-gravel percentage <20-40%	Structureless, mottled, burrowed, convoluted, parallel laminae; burrowed and parallel laminated limestone	Overall facies geometry—broad belts 20-40 miles (30-60 km) wide of low net and percentage sand-gravel between sand-rich fan lobes; interfan area broadens downdip and is similar to interchannel area; interfan is larger scale, broader than interchannel

ness, total sand and gravel content, and percentage of bed-load sediment. Only middle and distal fan facies are preserved in north Texas.

Each fan lobe displays similar lateral and down-fan facies patterns (Fig. G-5). Fan lobes are lobate sheets that become increasingly digitate at their distal margins. Lobes are separated by interfan deposits that are relatively poor in sand and gravel content. The Dalhart-Amarillo lobe is the oldest and largest of the Ogallala fan lobes in north Texas. Its features are representative of other fan lobes.

The Dalhart-Amarillo lobe consists largely of mid-fan deposits and covers approximately 15,000 sq mi (40,000 sq km). Mesozoic and latest Paleozoic strata underlie the deeply eroded pre-fan surface. Pre-fan topography was further modified in intrastratal salt solution and collapse.

The gradient of the fan surface, where least modified by postdepositional erosion, ranges from 20 to 25 ft/mi (4 to 5 m/km). Down-fan, the thick sand and gravel sheet evolves into digitate belts oriented northwest-southwest along the arc of the distal fan. Channel systems are 10 to 30 mi wide and contain 200 to 400 ft of accumulated sand and gravel in the middle fan. Distal-fan channel belts are separated by interchannel areas 5 to 20 mi wide that contain relatively little sand. Carbonate units, probably of lacustrine origin, are common in the muddy interchannel areas.

The Kern and laterally associated fans of the Great Valley, California were proposed as depositional analogs for the Ogallala by Johnson in 1901. The Kosi River fan, India, is a modern fan of comparable scale to a single Ogallala fan lobe. Similarities in interpreted

Table G-2. Comparison of depositional features of Ogallala and Kosi fans. (From Seni, 1980).

DEPOSITIONAL CHARACTERISTICS		KOSI RIVER FAN	SINGLE FAN LOBE CGALLALA FORMATION
FAN MORPHOLOGY		LOBATE GEOMETRY 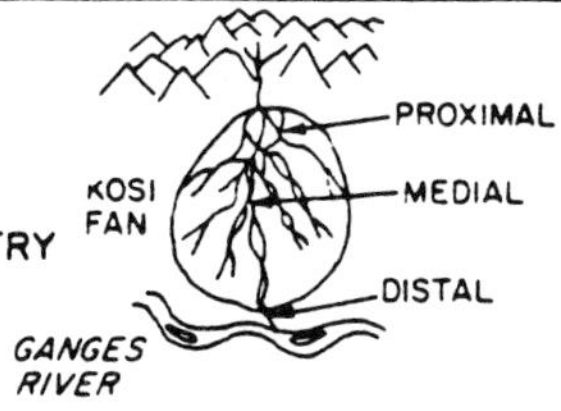 KOSI FAN; PROXIMAL; MEDIAL; DISTAL; *GANGES RIVER*	LOBATE TO DIGITATE GEOMETRY 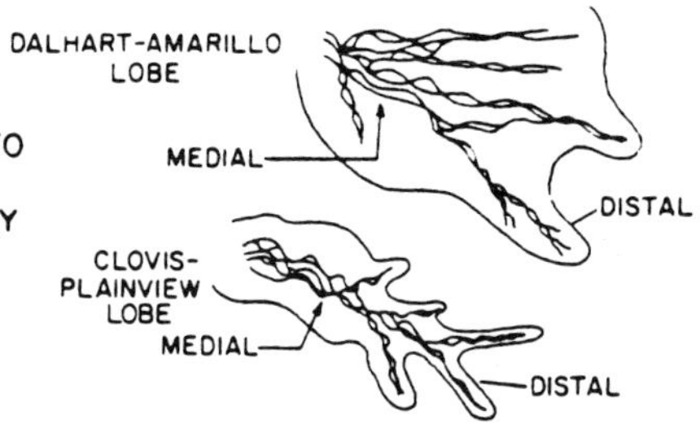 DALHART-AMARILLO LOBE; MEDIAL; DISTAL CLOVIS-PLAINVIEW LOBE; MEDIAL; DISTAL
SIZE		6000 mi^2 (15,000 km^2)	At least 5000 mi^2 – 15,000 mi^2 (13,000 to 40,000 km^2)
CHANNEL AND SAND-BODY MORPHOLOGY		SURFACE CHANNEL PATTERNS	SUBSURFACE SAND-BODY GEOMETRY
	PROXIMAL	SINGLE STRAIGHT CHANNEL LESS THAN 4 mi (6 km) WIDE	NOT PRESENT IN STUDY AREA
	MEDIAL	BRAIDED CHANNELS 4 to 10 mi (6-16 km) WIDE	BROAD SAND AND GRAVEL SHEET 25 to 100 mi (40 to 160 km) WIDE. INDIVIDUAL CHANNELS NOT RESOLVABLE
	DISTAL	SINGLE BRAIDED CHANNEL NARROWER THAN MEDIAL FAN	DIP-ORIENTED SAND-RICH BELTS BIFURCATING DOWNDIP 6 to 30 mi (10 to 50 km) WIDE
SLOPE GRAIN SIZE AND FAN FACIES		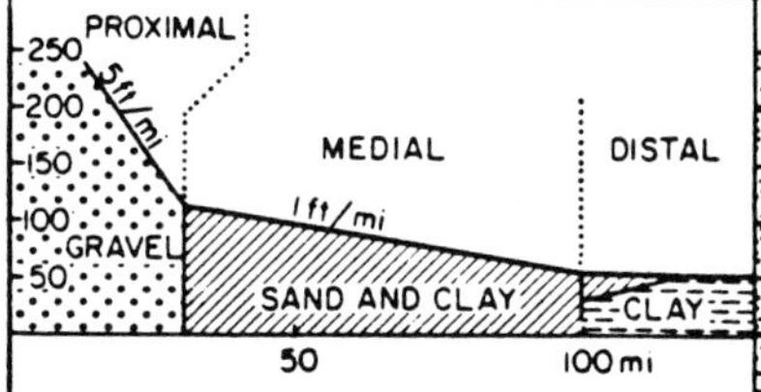 PROXIMAL; MEDIAL; DISTAL 250, 200, 150, 100, 50 5 ft/mi; 1 ft/mi GRAVEL; SAND AND CLAY; CLAY 50; 100 mi GRAIN SIZE BASED ON SEDIMENT LOAD OF KOSI RIVER. NO SUBSURFACE DATA	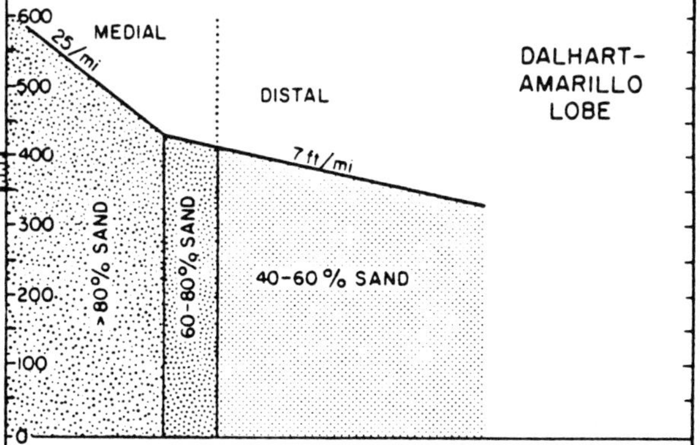 DALHART-AMARILLO LOBE MEDIAL; DISTAL 600, 500, 400, 300, 200, 100, 0 25 ft/mi; 7 ft/mi >80% SAND; 60-80% SAND; 40-60 % SAND GRAIN SIZE FROM SUBSURFACE DATA ONLY 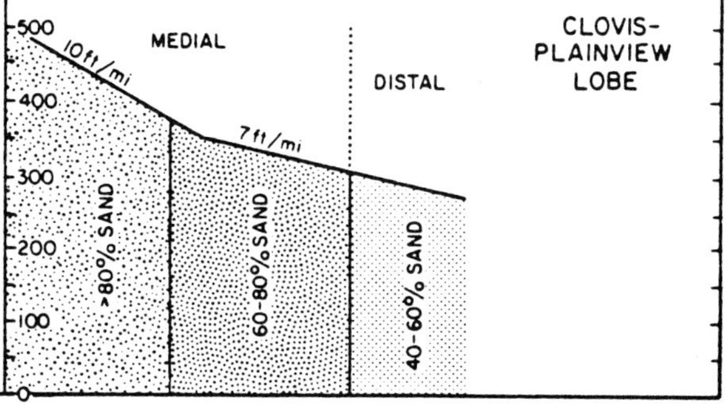 CLOVIS-PLAINVIEW LOBE MEDIAL; DISTAL 500, 400, 300, 200, 100, 0 10 ft/mi; 7 ft/mi >80% SAND; 60-80% SAND; 40-60% SAND

depositional style, geometry, size, stream gradients, composition, and sedimentary processes of the Kosi and Ogallala fans are shown in Table G-2.

Hawkesbury Sandstone, Sydney Basin

The Hawkesbury Sandstone (Triassic), which was described by Hobday and Jones (1982), conformably overlies fluvial and lacustrine facies of the Narrabeen Group over much of the Sydney Basin, Australia, (Fig. G-7) and is a classical example of a sandy, wet alluvial-fan system dominated by braided-stream processes. Along the basin flanks, the Hawkesbury is interpreted to lap out against older formations and against crystalline basement uplands. Channel and bar deposits are analogous to those of the modern Brahmaputra River, and the areal extent of the preserved system is some 20,000 sq km. Sandstone content exceeds 90 percent. Rather than a single large fan, the Hawkesbury probably consists of several overlapping fan lobes, each with its own point source located along the southwest margin of the basin. Its conformable relationship with the Narrabeen suggests that, as in some modern fans, the sandy, braided streamplain graded into mixed-load meanderbelt tracts separated by lacustrine floodbasins.

Bed-load channel facies evidence deposition in braided channels by large transverse or linguoid bars, sand waves, and subaqueous dunes. Channels had high width/depth ratios. During floods, thick but relatively lenticular, structureless mass-flow sands were deposited. Limited channel-margin accretion occurred along locally sinuous channel segments. Primary structures include giant crossbed sets (as much as 8 m thick) and a variety of smaller planar and trough crossbeds.

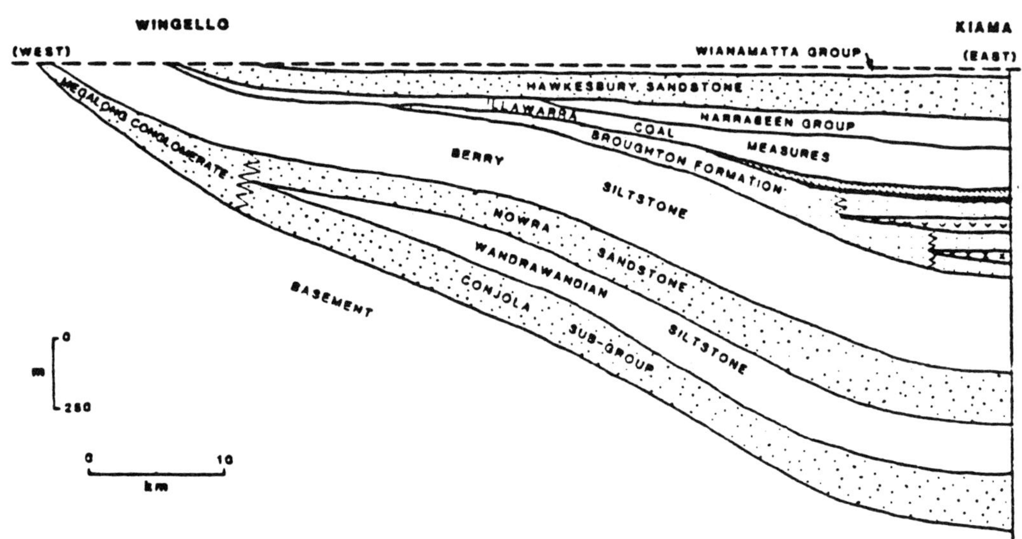

Figure G-7. Generalized east-west cross section through the southern Sydney Basin. (From Hobday and Jones, 1982).

Paleocurrent azimuths are consistent, with the vector mean at any locality varying within 25 degrees of the regional trend (Hobday and Jones, 1982).

Overbank facies consist of laterally persistent to lenticular siltstone and mudstone units. These fine-grained units are truncated by overlying channel fills and provide the source for the abundant mud clasts and blocks found within the Hawkesbury. Locally, medium- to coarse-grained sand beds and isolated dune trains record large-scale flooding into overbank areas. These finer grained units were deposited on active portions of the fan plain and in abandoned channel segments.

Genetic sequences. In parts of the Hawkesbury Sandstone, textural and structural sequences are random, producing a massive sand sheet (Fig. G-8). In other parts, well-developed vertical sequences include a basal scour surface overlain by pebbly sandstone which is in turn overlain by a progressively upward-fining package capped by siltstone. Cycles are 4 to 8 m thick and are commonly stacked. Such cycles are the product both of lateral accretion and of progressive channel abandonment.

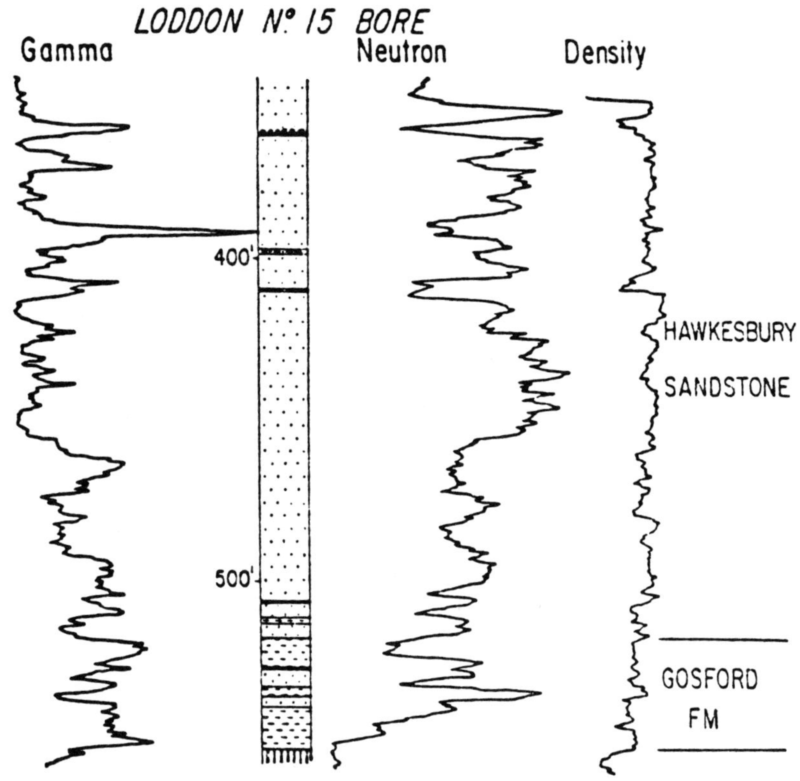

Figure G-8. Wire-line log typical of the Hawkesbury Sandstone and underlying Gosford Formation of the Narrabeen Group. (From Hobday and Jones, 1982).

CHAPTER 7

MEANDERING STREAMS -- MODERN AND ANCIENT

William E. Galloway

THE SPECTRUM OF FLUVIAL SYSTEMS

A fluvial system consists of a skeleton of fluvial channel-fill facies and closely associated splay and levee facies within a matrix of floodbasin muds and organics (Fig. G-9). Fluvial systems display a wide range of variation in such basic parameters as average proportion of sand to mud and dimensions and geometry of sand bodies. Consequently, they also vary in their capacity to transmit fluids. Further, within a fluvial system, systematic variations in these same parameters are observed to be both parallel and transverse to the sediment dispersal axis.

Large fluvial complexes tend to produce integrated drainage networks containing one or more trunk streams of the same type (i.e., meandering, braided, etc.) for large parts of the network. Depositional characteristics of these trunk streams provide a logical basis for differentiating significantly different portions of a fluvial system. In a series of papers, Schumm (1960, 1972) has described relationships in modern streams among sediment load transported by the channel, channel geometry, and sediment type deposited by the channel. These relationships can be quantified for modern river segments (Table G-3) and provide qualitative trends that can be applied in interpretation and classification of ancient fluvial depositional systems.

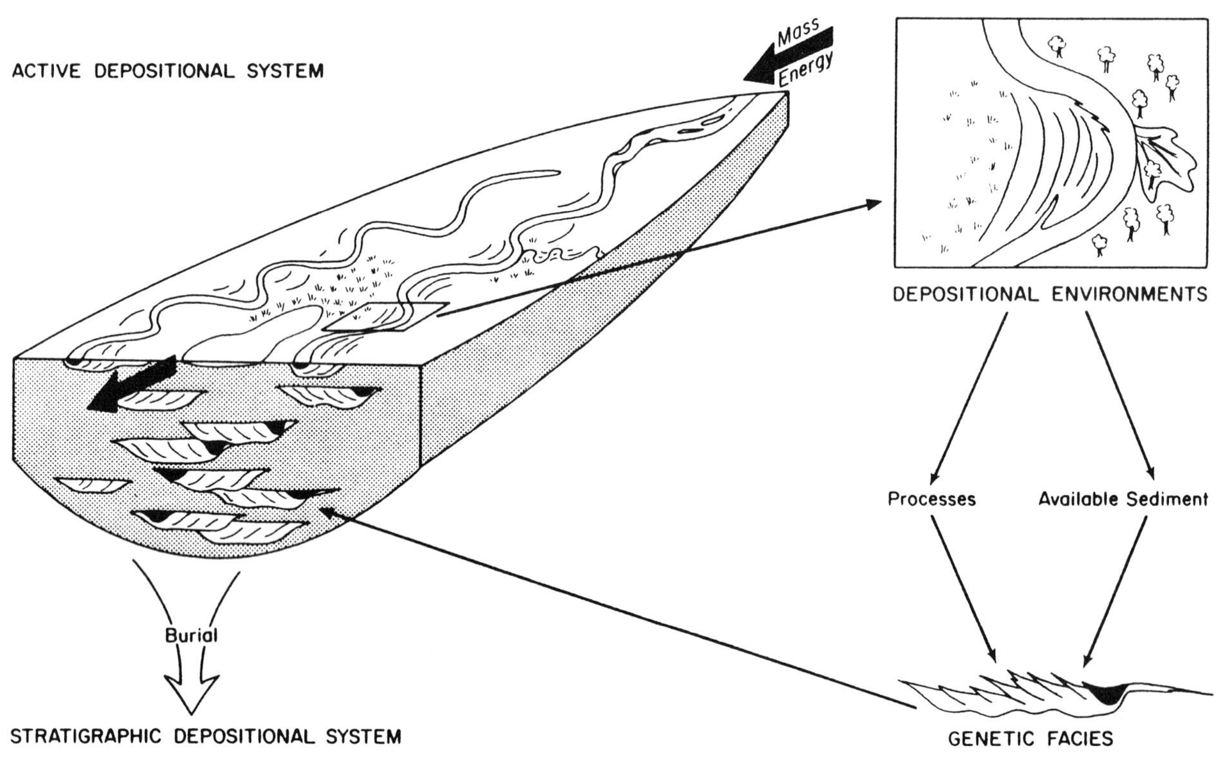

Figure G-9. The environmental and genetic facies framework of a fluvial depositional system. (From Galloway and others, 1979).

The basis of Schumm's classification lies in the empirical observation of a fundamental correlation between the ratio of bed load to suspended load transported by a stream and the cross-sectional geometry of the channel (expressed as width/depth ratio). This relationship is independent of other variables, such as slope, discharge, or periodicity of flow. Thus, alluvial channels can be classed as bed-load, mixed-load, or suspended-load types, depending on the sediment load transported by the channel. Furthermore, for each channel type, modern streams are characterized by consistent patterns of erosion and deposition, ranges of sinuosity, and proportion of suspended load deposited within the channel perimeter.

Bed-load and suspended-load channels can be considered end members of a spectrum of possible fluvial channel types into which any channel segment can be fit. At one end of the spectrum are suspended-load-dominated channels, characterized by low width/depth ratios, high sinuosity, dominance of vertical erosion and deposition, and a relatively fine grained channel fill containing abundant suspended-load sediment. At the other end are bed-load-dominated channels, characterized by very high width/depth ratios, straight to slightly sinuous channel patterns, lateral erosion and deposition, and coarse-grained channel fill containing little suspended-load material. Mixed-load channels occupy the middle range of the spectrum. Further, bed-load channels typically have steep gradients, and suspended-load channels have lower gradients. Thus, a commonly observed decrease in stream gradient along its length is paralleled by a systematic, downchannel change in channel geometry and types of fluvial deposits.

Absolute dimensions of the channel (and its resultant channel-fill deposits) are determined by stream discharge. Because the average

Table G-3. Classification of alluvial channels. (From Schumm, 1972).

Mode of Sediment Transport and Type of Channel	% Silt and Clay Deposited in Channel Perimeter (M)	Bed-Load (Percentage of Total Load)	Channel Stability		
			Stable (Graded Stream)	Depositing (Excess Load)	Eroding (Deficiency of Load)
Suspended Load	>20	<3	Stable suspended-load channel. Width-depth ratio <10; sinuosity usually >2.0; gradient relatively gentle.	Depositing suspended load channel. Major deposition on banks causes narrowing of channel; initial streambed deposition minor.	Eroding suspended-load channel. Streambed erosion predominant; initial channel widening minor.
Mixed Load	5-20	3-11	Stable mixed-load channel. Width-depth ratio >10, <40; sinuosity usually <2.0, >1.3; gradient moderate.	Depositing mixed-load channel. Initial major deposition on banks followed by streambed deposition.	Eroding mixed-load channel. Initial streambed erosion followed by channel widening.
Bed Load	<5	>11	Stable bed-load channel. Width-depth ratio >40; sinuosity usually <1.3; gradient relatively steep.	Depositing bed-load channel. Streambed deposition and island formation.	Eroding bed-load channel. Little streambed erosion; channel widening predominant.

width/depth ratio of bed-load channels has been found to exceed 40:1, bedload channels 2 ft and 20 ft deep will be at least 40 times greater, or more than 80 ft, or 800 ft wide, respectively.

Additional factors influence deposition within each of the broad classes of channel type (see Miall, this volume). Climate affects the flow periodicity and the presence (and type) of vegetation. Vegetation itself is a factor, as it stabilizes channel banks and reduces bank erosion, thus favoring bed erosion during rising water levels. Despite these other variables, considerable data from modern channels support the basic relationships between sediment transport, sediment deposition, and channel geometry described by Schumm.

INTERPRETATION OF PALEOCHANNELS

Determination of cross-sectional dimensions or original bed-load/suspended-load ratio of a paleochannel is extremely difficult even under optimum conditions of exposure. However, channel-fill composition, geometry, and internal structure provide a basis for modifying Schumm's classification of modern channels for application to the stratigraphic record. The modified classification utilizes fundamental attributes of any ancient depositional system--composition and geometry of the framework elements--that can be defined from outcrop or subsurface data, or both (and that are basic controls of postdepositional ground-water flux and associated epigenetic mineralization processes). Well-known fluvial models can be placed within a spectrum of fluvial systems analogous to that described by Schumm's classification. Importantly, fluvial sequences that do not conform to idealized models can also be interpreted and placed within a predictive framework. The composition and geometry of bed-load, mixed-load, and suspended-load paleochannel fills are discussed below and are summarized in Figure G-10.

Bed-load Channel

The channel-fill sequence is dominated by sand. Coarse sand and gravel are commonly present but are not necessary components; the sequence could consist exclusively of fine to medium sand. A bed-load channel is characterized by a high width/depth ratio and a tendency to erode laterally; these factors produce a tabular or belted sand body. Development of multilateral channel fills may produce a fluvial sheet sand. Straight channels are characterized by relatively uniform depths of scour along the base, and this may be reflected in the low relief on the base of the sand body and by preservation of laterally continuous sheet or tabular remnant floodplain mud units. Amalgamated, multilateral sand bodies result from rapid lateral migration of bed-load channels (Fig. G-10). Internally, bed-load channel-fill sequences reflect the dominance of bed accretion units, such as longitudinal, transverse, lateral, and chute bars, which produce a channel fill consisting of multiple, interlensed depositional units of varying grain size and displaying complex textural sequences. Braided and coarse-grained meanderbelt models are well documented types of bed-load channel-fill sequences (Fig. G-10).

Mixed-load Channel

Mixed-load paleochannel deposits consist of sand with subordinate silt and clay. That channels were sinuous, with variable depths of

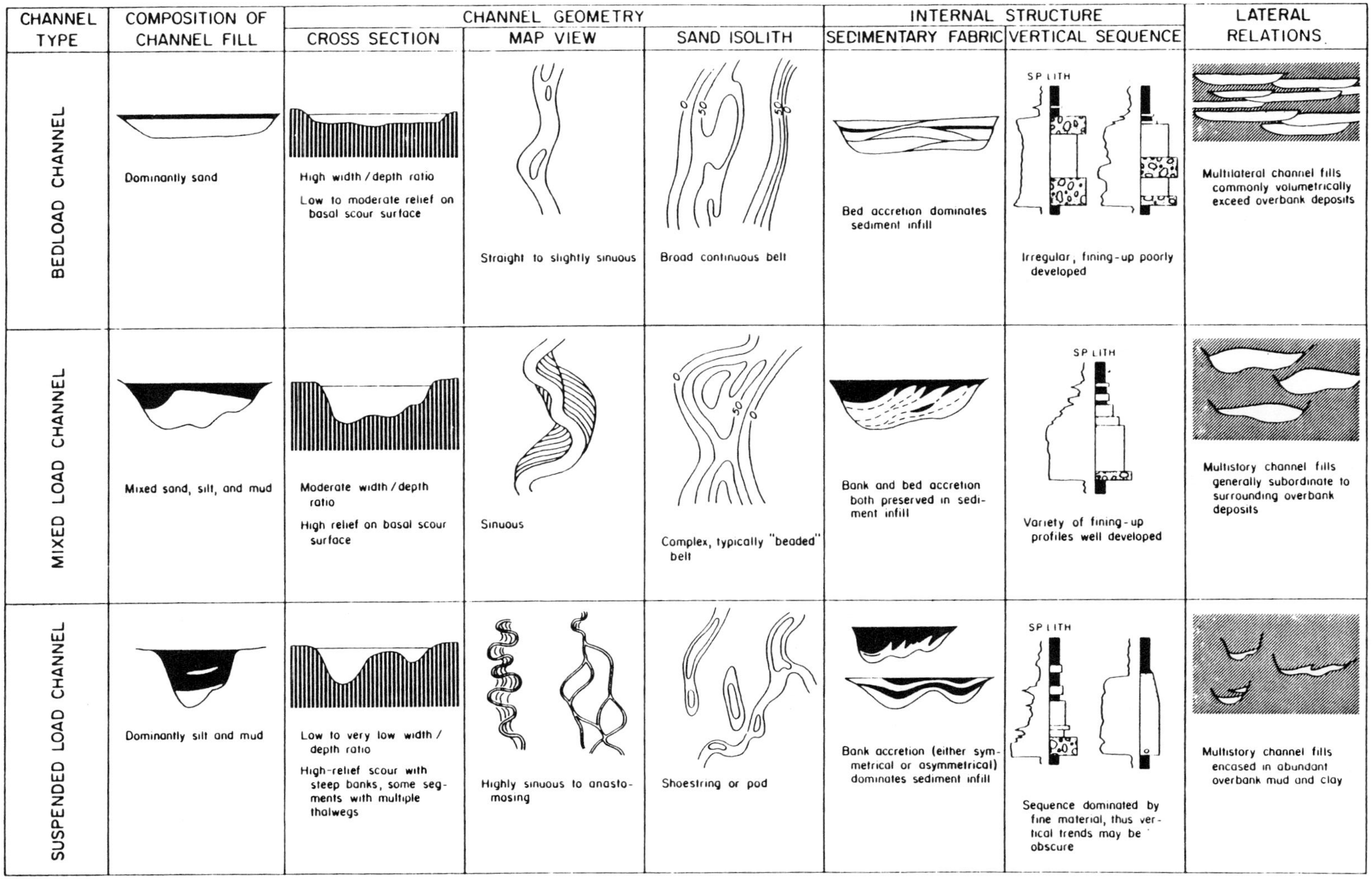

CHANNEL TYPE	COMPOSITION OF CHANNEL FILL	CHANNEL GEOMETRY			INTERNAL STRUCTURE		LATERAL RELATIONS
		CROSS SECTION	MAP VIEW	SAND ISOLITH	SEDIMENTARY FABRIC	VERTICAL SEQUENCE	
BEDLOAD CHANNEL	Dominantly sand	High width / depth ratio Low to moderate relief on basal scour surface	Straight to slightly sinuous	Broad continuous belt	Bed accretion dominates sediment infill	SP LITH Irregular, fining-up poorly developed	Multilateral channel fills commonly volumetrically exceed overbank deposits
MIXED LOAD CHANNEL	Mixed sand, silt, and mud	Moderate width / depth ratio High relief on basal scour surface	Sinuous	Complex, typically "beaded" belt	Bank and bed accretion both preserved in sediment infill	SP LITH Variety of fining-up profiles well developed	Multistory channel fills generally subordinate to surrounding overbank deposits
SUSPENDED LOAD CHANNEL	Dominantly silt and mud	Low to very low width / depth ratio High-relief scour with steep banks, some segments with multiple thalwegs	Highly sinuous to anastomosing	Shoestring or pod	Bank accretion (either symmetrical or asymmetrical) dominates sediment infill	SP LITH Sequence dominated by fine material, thus vertical trends may be obscure	Multistory channel fills encased in abundant overbank mud and clay

Figure G-10. Geomorphic and sedimentary characteristics of bed-load, mixed-load, and suspended-load channel segments. (From Galloway, 1977).

basal scour along the thalweg, is reflected in the development of irregular or beaded belts of sand (Fig. G-10) and in the preservation of lenticular, discontinuous floodbasin remnants between channel-fill sequences. A record of mixed bank- and bed-accretion (point-bar and channel-lag) deposits is preserved in the channel fill, and a repetitive upward-fining sequence typifies most vertical sections through the meanderbelt deposit. Channel meandering and consequent point-bar accretion, combined with the stacking and amalgamation of successive channel fills, commonly produces a composite sand body that is much larger than the original channel. The channel plug provides the best record of true channel dimensions.

Suspended-load Channel

Suspended-load paleochannels were narrow and confined; erosion occurred primarily at the base, and surrounding clay-rich sediments formed stable, steep channel banks. Deposition of channel fill was dominantly along the banks and may have been one-sided in highly sinuous channel segments or symmetrical in straight channel segments (Fig. G-10). The dominant channel-fill sediment ranges from very fine sand to silt and clay, but some very coarse sediment may be present. Sand-body geometry is highly lenticular in cross section and forms observable sinuous or anastomosing patterns (Fig. G-10). Channel-fill units are typically encased in fine-grained floodbasin deposits and tend to stack vertically. Vertical sequences may fine upward or may show little variation if the range of grain sizes is limited.

The distributary channel model provides an example of a straight, suspended-load channel type. Point bars of the Mississippi River upper delta plain record deposition by a sinuous suspended-load channel.

INTERPRETATION OF FLUVIAL SYSTEMS

A fluvial system, or major segments of a fluvial system, can be categorized in terms of the dominant channel type (bed load, mixed load, or suspended load) of the trunk stream(s). Thus, classification is based on geometric and compositional attributes of the framework sands, which, in turn, reflect important characteristics of the system as a potential aquifer. It is important to note that a wide range of channel types commonly coexists within a single fluvial system. For example, crevasse channels flanking the main channel will likely be enriched in suspended load. Tributaries can range from bed- to suspended-load type channels, depending on their provenance and gradient. The main channel itself may vary in type where it is locally affected by tributary input or incision into unusually sandy or muddy substrates. However, the dominant channel type of the trunk streams is a first-order basis for interpretation and categorization of the system.

As pointed out by Brown and others (1973) and Schumm (1977), an ideal fluvial system displays a systematic downflow evolution of process, of morphology, and of depositional facies. A hypothetical example of a complete system is illustrated in Figure G-11. This schematic fluvial system traverses two major depositional basins. The first is a montane basin opening into an aggrading riverine plain, such as the Murrimbidgee Riverine Plain of New South Wales, Australia, described by Schumm (1977). The second depositional basin is a progradational coastal plain bordering an ocean. The river begins as a network of bed-load tributaries, undergoes considerable metamorphosis

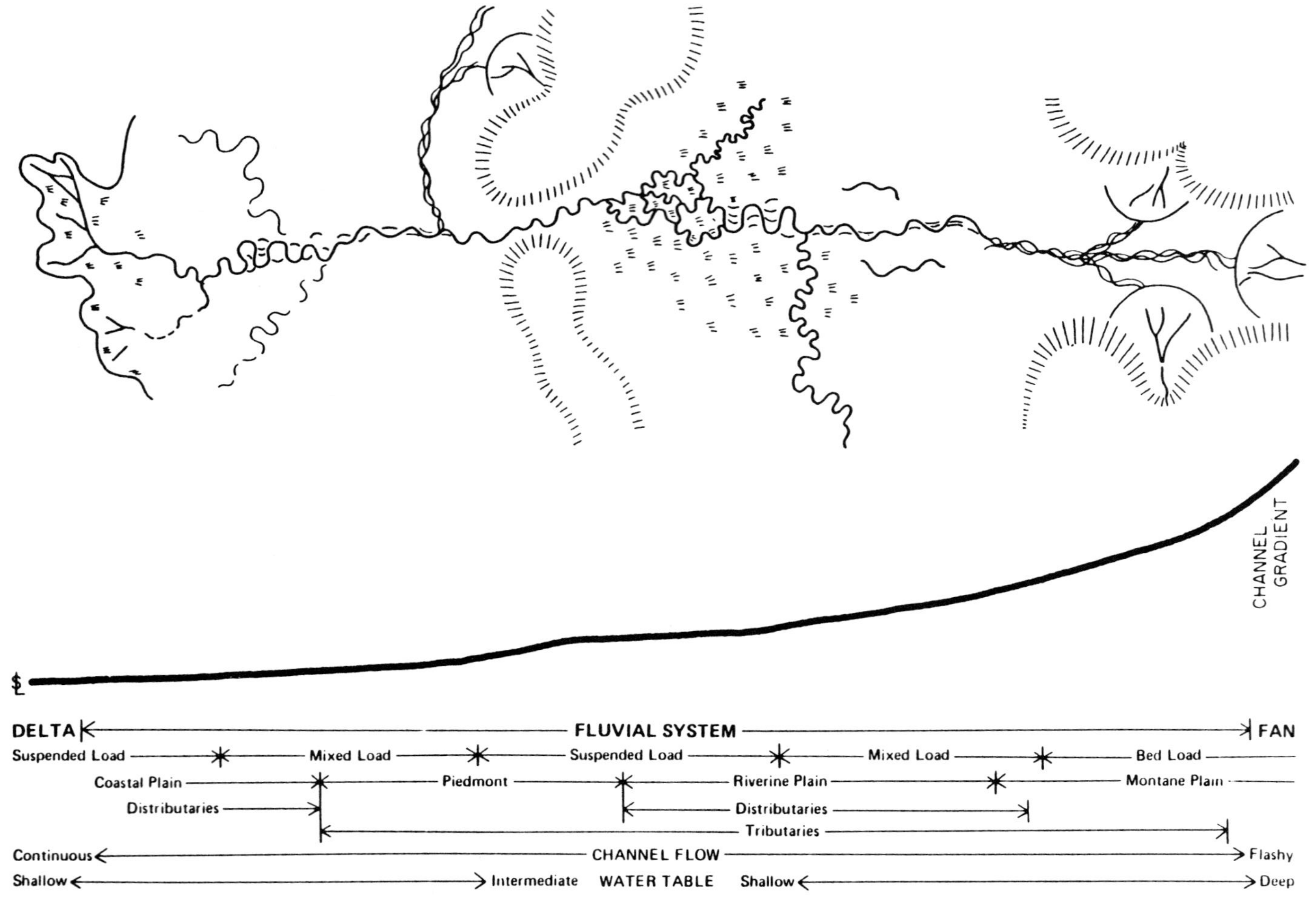

Figure G-11. Hypothetical fluvial system traversing a montane basin and associated riverine plain and debouching across a piedmont into a prograding, deltaic coastal plain. The trunk channel evolves through bedload, mixed-load, and suspended-load phases as gradient, sediment load, and discharge vary along the system. (From Galloway and Hobday, 1983).

in response to down-channel changes in gradient, sediment supply, and flow characteristics, and ends as a suspended-load distributary channel complex on a delta plain. The system illustrated may be considered "complete" in that the entire range of possible channel types is present. It is equally possible that the fluvial system might reach its final depositional basin (be it lacustrine, inland sea, or oceanic) much earlier in its evolution. If the provenance of the system provides dominantly suspended load to the tributaries, bed-load channels will be absent. Generally, however, large integrated fluvial systems will display a broad spectrum of channel types along their course.

In addition to differing channel-fill facies, ancient bed-load, mixed-load, and suspended-load fluvial systems (Fig. G-12) each display characteristic facies associations.

Bed-load fluvial systems consist dominantly of channel and channel-flank deposits; floodbasin facies are subordinate and consist of floodplain sandy mud and silt. Associated eolian deposits are common, even in subhumid climates, because broad expanses of unconsolidated sand are exposed to wind reworking during low water levels. The water table in interchannel areas is typically deep, because such systems occur in the upper reaches of the sedimentary basin, and the sandy floodplain and channel sediments are highly permeable. Crevasse splays may be well developed and contain significant amounts of bed-load sediment. Modern examples are the Brahmaputra River (Coleman, 1969) and the Platte River (Smith, 1970).

Mixed-load fluvial systems typically preserve a high percentage of floodbasin deposits, including floodplain muds and silts and backswamp carbonaceous muds and clays. Sand percentages may be high along principal fluvial axes where channel-fill units are vertically stacked and

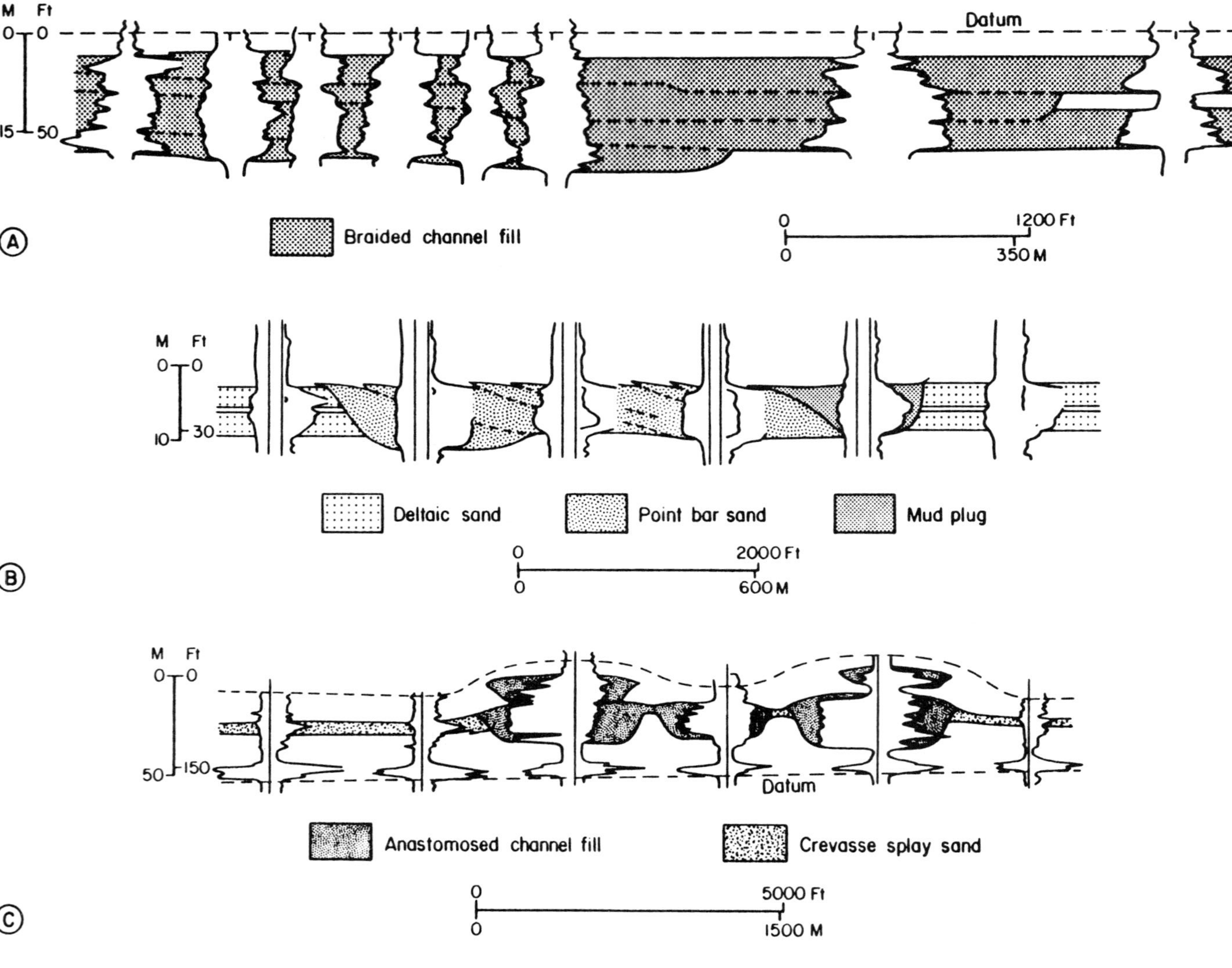

Figure G-12. Channel-fill sand bodies of a braided bed-load channel (A), meandering mixed-load channel (B), and anastomosing, low-sinuosity, suspended-load channel complex (C) as interpreted from electric logs. A. Uranium-bearing Oakville Sandstone (Miocene), Texas Gulf Coast Basin, U.S.A. B. Oil-bearing Fall River Sandstone (Cretaceous), Powder River Basin, U.S.A. C. Petroleum-bearing Cantuar Formation of the Mannville Group (Cretaceous), Saskatchewan, Canada. (From Galloway and Hobday, 1983).

amalgamated; however, overall sand percentage averages 20 to 40 percent. Channel fills are flanked by crevasse-splay and levee sands and silts. The water table is usually defined by the water level in the main channels; thus levee and crevasse deposits are typically leached and oxidized. Much of the channel fill remains saturated, and organic debris deposited within the lower point bar and channel lag is commonly preserved. Broad, low-lying backswamps also have shallow water tables, and organic debris and syngenetic sulfides are commonly preserved within the mud and clay deposited in this environment. Modern examples are the lower Brazos River (Bernard and others, 1970) and the Mississippi River (Fisk, 1944).

Suspended-load fluvial systems have low sand percentages. Flood-basin deposits consist dominantly of backswamp and lacustrine muds and clays, and levee deposits are well developed. Crevasse splays may be common but consist almost entirely of suspended-load sediment. Sand isolith maps show complex anastomosing or distributary patterns. The low topographic gradients and relatively impermeable sediments typical of suspended-load fluvial systems result in generally shallow to very shallow water tables that intersect the land surface along channels and in backswamp and lacustrine basins. Considerable organic debris and syngenetic sulfides are commonly preserved in such a system. Modern examples are the lower Mississippi and the Atchafalaya Rivers (Fisk, 1947, 1952; Frazier, 1967).

MEANDERING RIVERS I: DEPOSITIONAL ELEMENTS

Moderate- to high-sinuosity channels include many depositional environments and corresponding depositional architectural elements.

Channel Lag

Channel lag consists of the coarsest sediment transported by the river, locally derived debris such as mud clasts and blocks eroded from the banks and bottom, and water-logged debris. The lag forms the channel floor (Fig. G-13) and varies in grain size and composition, depending on local depth of scour and location within the channel segment. Coarsest lag is found in areas of maximum velocity and turbulence.

Point Bar

Sediment eroded from upstream channel banks is transported up gently sloping convex stream banks into areas of relatively low velocity and turbulence where it is deposited on laterally accreting point bars (Fig. G-13). Cross-sectional area of the channel is maintained by concomitant erosion of the convex, or cut bank. Thus, the curvature of the meander tends to become increasingly exaggerated. Because sediment is moving up and out of the channel onto the bar, a vertical decrease in grain size characterizes the point-bar sequence. The accretionary architecture of the point bar is commonly reflected by the development of ridge-and-swale topography (also known as meander scrolls) on the point-bar surface (Fig. G-13). Fine sediment washed across the bar surface during floods may pond in the swales, forming local muddy lenses and plugs. Older portions of the point bar are commonly vegetated and capped by fine-grained levee and floodplain sediment, completing the upward-fining cycle that begins with the channel lag. Sand transport across lower- and mid-bar surfaces is dominantly by dune migration, and medium-to-large trough cross-stratification

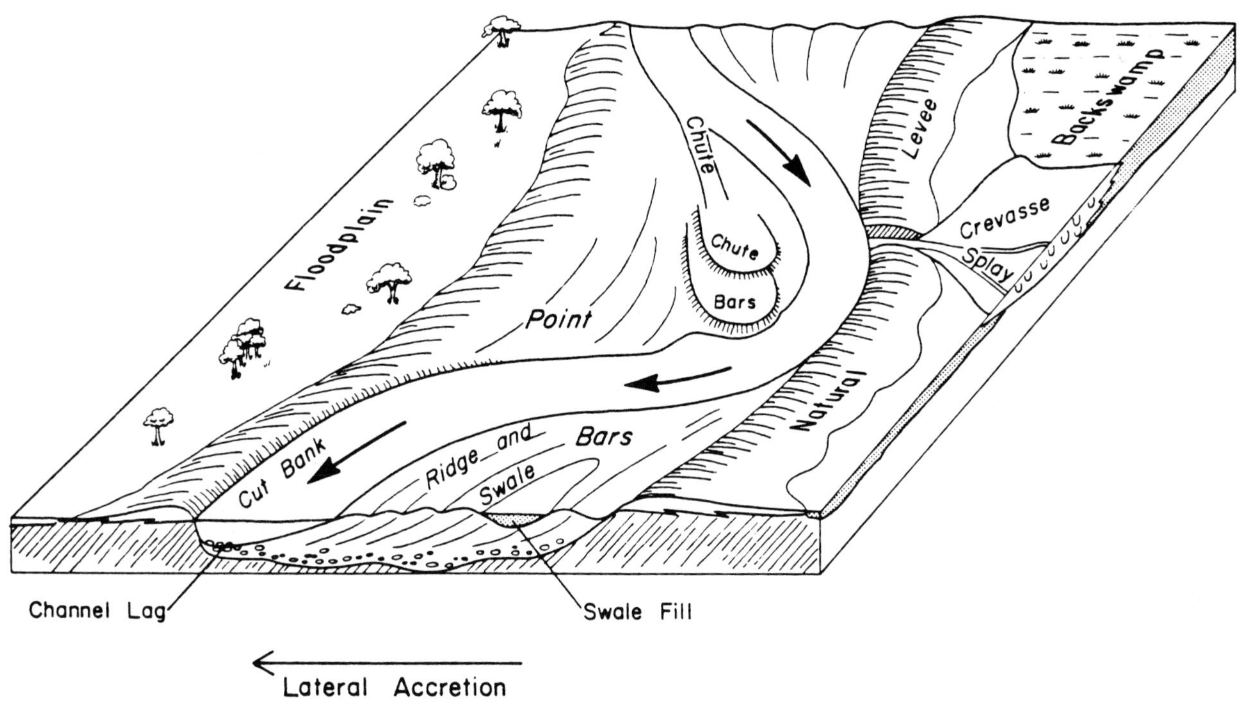

Figure G-13. Depositional environments and physiographic features of a sinuous channel segment. (From Galloway and others, 1979).

characterizes this part of the sand body. Ripple, climbing-ripple, tabular and planar stratification occur in the finer grained upper point-bar sequence where water depths are shallow, or flow velocities are lower, or both. The bar surface may be modified by sheetwash and gullying as well as by rooting and burrowing during periods of subaerial exposure.

Chute and Chute Bar

During flood stage, part of the river flow may cut directly across the surface of the point bar. When this happens, significant amounts of coarse bed-load sediment move up out of the main channel and are funneled through one or more chute channels that have been scoured into the upstream end of the point bar. As flow spreads across the bar surface, bed-load material is deposited, forming a chute bar. The processes, physiography, and sediments of chute-modified, coarse-grained point bars have been described by McGowen and Garner (1970), Levey (1978), and Arche (1983).

The chute contains coarse lag material typically found in the main channel. Chute bar units are thickest where the flow reenters the main channel and the bar progrades into the relatively deep water of the channel. Chute bars consist of relatively coarse sediment deposited by flow separation at the lee side of the bar crest. Typical structures of the chute complex include imbricated pebble sheets and lags in upstream parts of the chute channel, trough (dune) and ripple lamination in medial and distal portions of the chute, and planar and avalanche cross-stratification in the chute bars. Cut-and-fill associated with scour around vegetation or other obstructions to flow is also a common feature of the chute-modified point bar. The principal result of the chute development is the deposition of coarse bed-load sediment and large-scale sedimentary structures at the top of the point-bar sequence.

Channel Plug

Sudden abandonment of a channel segment by upstream channel piracy, avulsion, or meander cutoff terminates the further influx of bed-load sediment. The channel is slowly filled by suspended-load material washed in across the floodplain from another river segment. While being filled, the channel may form a local lake (oxbow lake) or swamp. Consequently, organic debris or even lignite may constitute an important component of the abandoned-channel plug. Muddy channel plugs provide a true picture of actual channel dimensions; thus the plug is narrow (tens to a few hundred feet) and elongate. In complex meanderbelt sequences, channel plugs intricately compartmentalize upper portions of otherwise laterally extensive meanderbelt sand bodies.

MEANDERING RIVERS II: FACIES MODELS

Two significantly different models of sinuous channel-fill sequences have been summarized from a considerable number of studies of modern fluvial systems. It should be emphasized that these models are distillations of common features seen in many modern rivers; consequently, actual channel-fill sequences may be expected to vary from the idealized models. As pointed out by Jackson (1975), considerable variation in composition, vertical sequence, and internal structure exists within a single point bar. Nonetheless, the fine-grained point-bar model and the coarse-grained, chute-modified point-bar model summarize salient attributes of many sinuous channel-fill sequences. The fine-grained meanderbelt model (Fig. G-14) is the best known fluvial model. It characterizes most highly meandering channel systems and is readily recognized by its highly ordered internal architecture.

Geometry: Belts or "beaded" belts of lenticular, erosionally based sand are dip oriented. The upper boundary is commonly transitional into overlying sediments; multistory and vertically amalgamated sand bodies are common. Width/thickness ratios of sand bodies are moderate to high. Tributary and distributary sand patterns are common.

Composition: The composition is chiefly of very fine to medium sand, but a broad range of grain sizes, including gravel and mud, occurs in the complete sequence. Mud clasts and blocks and woody plant debris are concentrated near the base.

Depositional Units: These consist of channel-lag, point-bar (composed of a succession of lateral-accretion units), and top-stratum deposits. Fine-grained channel plugs record final filling of the channel, and thinner swale-fill lenses locally cap the point-bar units.

Internal Structures: Sands are characterized by medium- to large-scale trough cross-stratification, with local zones of tabular, planar, and ripple stratification. The finer sediments at the top of the sequence commonly contain abundant small-scale structures, including ripple and ripple-drift laminations, mud drapes, and root traces. The average paleocurrent vector of the directional features parallels the sand-body trend, but considerable variability exists locally.

Vertical Sequence: This is characterized by a general upward decrease in both grain size and scale of sedimentary structures. Ver-

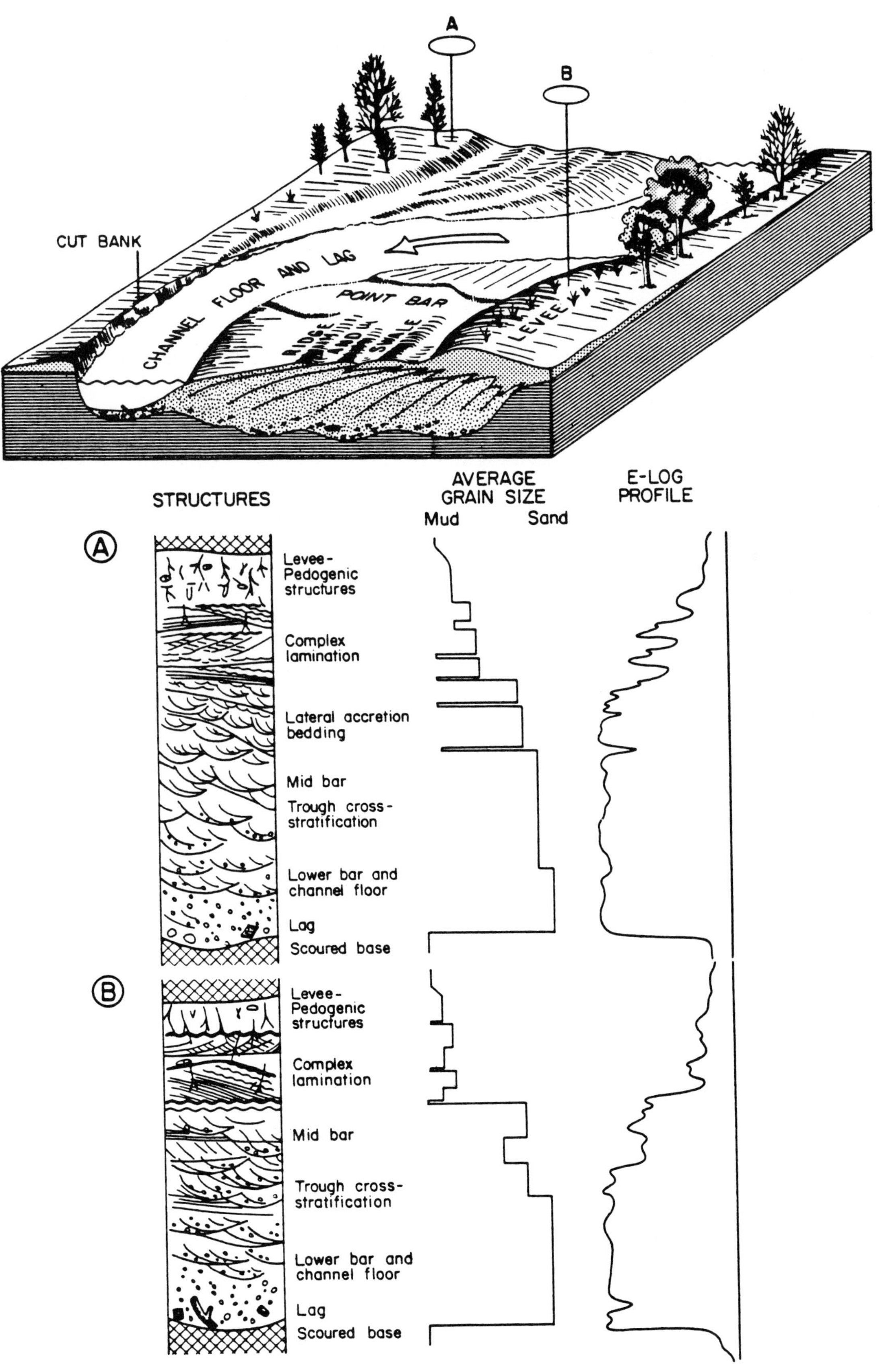

Figure G-14. Point-bar model. (From Galloway and Hobday, 1983).

tical trends may be gradational or quite erratic in detail, depending on the flow variability of the original river, the range of sediment sizes available, or other factors.

The coarse-grained, chute-modified meanderbelt model (Fig. G-15) illus-trates significant differences in the channel-fill sequence produced by the development of chutes and chute bars.

Geometry: Broad, laterally and vertically amalgamated, and generally dip-oriented belts of sand occur. The width/thickness ratio of the sand body is moderate to high, and development of coalesced multilateral channel fills is common.

Composition: Channel fill is composed chiefly of fine to coarse sand; gravel zones are common, especially near the base and top of the sequence. Mud clasts are large; fragmental to finely macerated plant debris are common minor constituents. Fine-grained sediments are limited to the thin top-stratum and channel-plug deposits.

Depositional Units: Channel lag is overlain by lateral accretionary chute-bar sands deposited by migrating subaqueous dunes. The upper bar sequence consists of interlensing scour-based chute-fill and accretionary chute-bar sands.

Internal Structures: Large- to medium-scale trough cross-stratification with some tabular cross-stratification dominates the lower portion of the channel fill. Large-scale avalanche cross-stratification as well as trough, scour-and-fill, discontinuous gravel lags, and minor planar and ripple laminations characterize the upper channel-fill sequence. The average vector of paleocurrent indicators parallels the channel trend, but considerable local variability results from the complex depositional patterns.

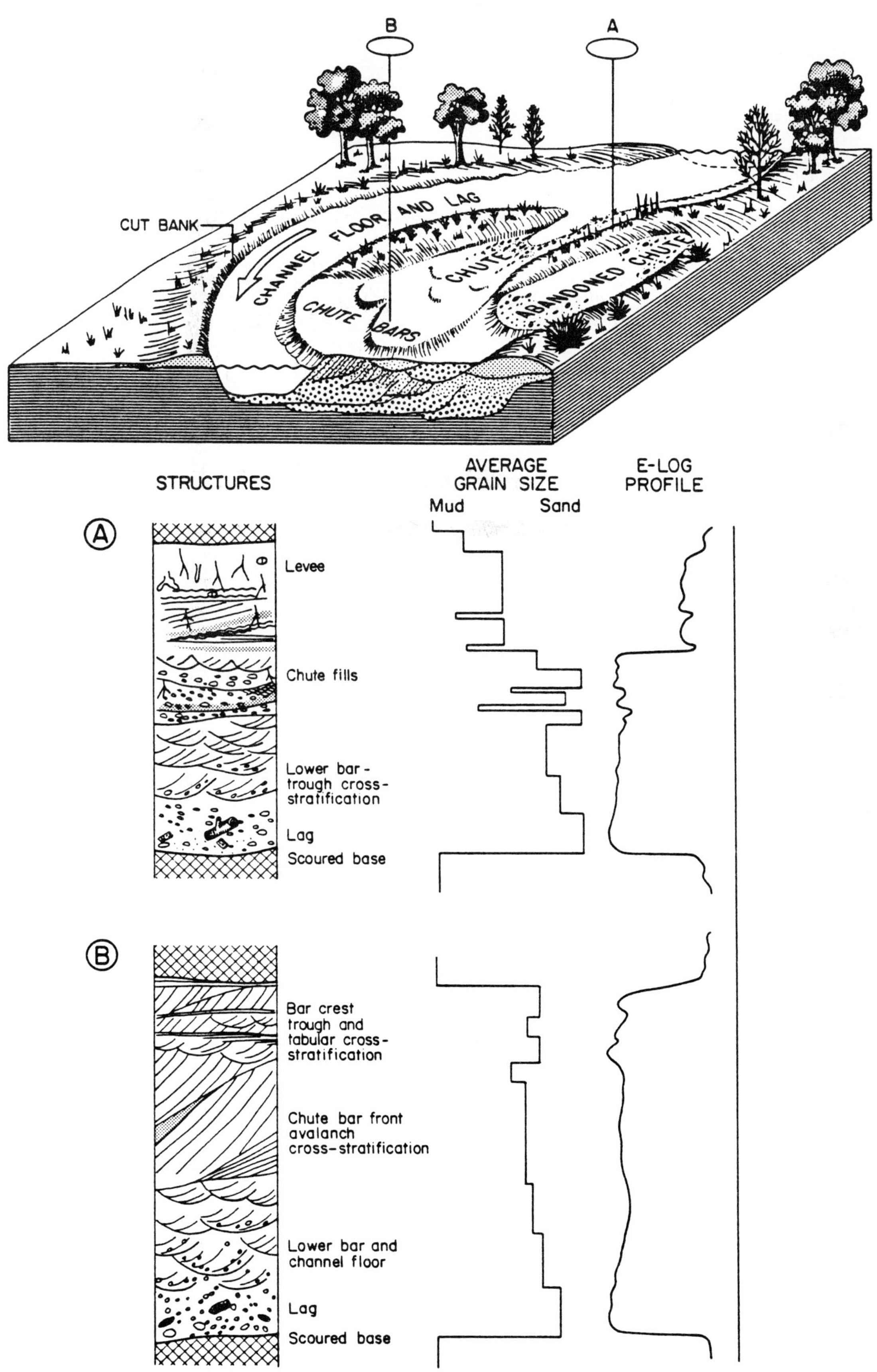

Figure G-15. Coarse-grained, chute-modified point-bar model. (From Galloway and Hobday, 1983).

Vertical Trends: In contrast to the trend characteristic of the fine-grained meanderbelt, grain size exhibits little vertical fining. In fact, some of the coarsest sediment and largest structures may occur at the top of the sand body in chute and chute-bar units.

CHAPTER 8

COAL DEPOSITS IN CRETACEOUS AND TERTIARY FLUVIAL SYSTEMS OF THE ROCKY MOUNTAIN REGION

Romeo M. Flores
U.S. Geological Survey
Denver, Colorado 80225

INTRODUCTION

The exploitation of coal deposits, whether in strip mines or deep mines, is dependent upon discrete coal bodies that are thick in some places, and thin or absent in others. Over the past few decades, depositional modeling has made possible a greater understanding of the physical characteristics of coal bodies and can be used to explain (or predict) where they are thick, thin, or absent.

Depositional-environment models now available are sufficiently refined so that they can assist not only in solving problems in coal exploration and development, but also in mapping hydrocarbon reservoir trends. The usefulness of the concept of coal depositional modeling to petroleum exploration is based on the premise that coal bodies were deposited in an environment juxtaposed to potential hydrocarbon reservoir bodies (e.g., channel sandstone, crevasse-splay sandstone). Thus, the distribution and geometry of coal bodies can be used to predict, identify, and delineate reservoir bodies. Coal is a readily identifiable lithotype in geophysical logs, and its signature is not adversely affected by diagenesis. Because of this advantageous attribute and because of the mutual environmental association of coal and reservoir rocks, coal can be a useful tool in facies-oriented exploration. In order to better understand this method of exploration, attributes of Cretaceous and Tertiary coal-bearing deposits in fluvial systems are examined.

Cretaceous and Tertiary coals were deposited in three major environments in fluvial systems in the Rocky Mountain region: lower alluvial plain of a coastal plain-piedmont setting; upper alluvial plain of a coastal plain-piedmont setting; and intermontane alluvial plain. The following discussions will demonstrate facies characteristics, sequences, and associations in the Cretaceous Crevasse Canyon and Menefee Formations in the Gallup sag, New Mexico; the Cretaceous and Tertiary Raton and Poison Canyon Formations in the Raton Basin, Colorado and New Mexico; and the Tertiary Wasatch Formation in the Powder River Basin, Wyoming (Fig. 1). Coal deposition in the lower alluvial plain is typified by the Gallup sag deposits, in the upper alluvial plain by the Raton Basin deposits, and in the intermontane alluvial plain by the Powder River Basin deposits.

GENETIC (FLUVIAL) FACIES

Facies of fluvial systems, like those of other depositional systems, may be subdivided into framework and nonframework facies. The framework facies include the skeletal coarse-grained channel-levee facies. This facies is bounded by the nonframework, fine-grained floodplain (crevasse splay and lacustrine) and backswamp facies. In the three case study areas, these facies display a wide range of variation in geometry and arrangement or architecture. Important factors that control architecture of fluvial facies in these areas are eustatic rise of sea level and extrabasinal and intrabasinal tectonism. These allocyclic factors determine the vertical organization of sedimentary facies, styles of lateral facies distribution, and accumulation of organic deposits.

Channel-Levee Facies

The channel-levee facies consist mainly of channel sandstone bounded by levee deposits of silty sandstone, siltstone, and shale. These facies occur as either isolated or clustered deposits. The channel sandstones have erosional bases, are fine to coarse grained, and form a fining-upward

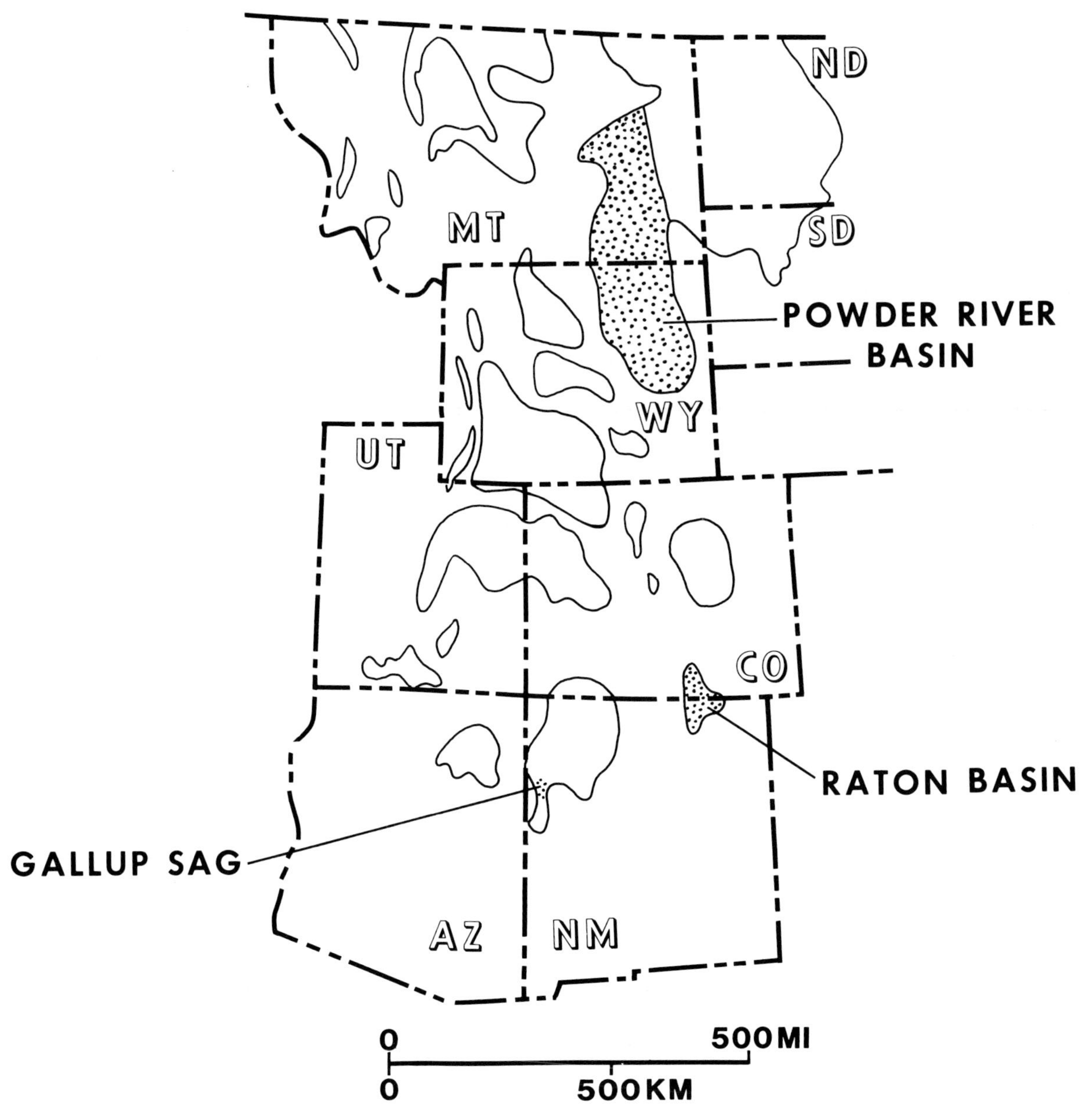

Figure 1. Map of the Rocky Mountain coal basins showing locations of Gallup sag, Raton Basin, and Powder River Basin.

sequence. Some channel sandstones consist of more than one body and include accretion bar deposits separated by epsilon crossbeds (Allen, 1963), indicative of lateral migration of the channels. Other channel sandstones, which do not show accretion units, suggest deposition in low-sinuousity stream channels. The lower part of a typical channel sandstone contains large-scale trough and planar crossbeds locally interrupted by large convolute laminations; the upper part contains small-scale trough crossbeds, convolute lamination, and ripple lamination. The large-scale crossbedded lower part probably represents dune and bar migration in the main channel fill. The small-scale crossbedded upper part probably represents channel fill in shallower water during abandonment. The channel sandstones contain lag deposits of rounded sandstone, siltstone, shale, limestone, ironstone, carbonaceous shale, coal spar, and mollusk shell fragments, which probably represent reworked deposits of the floodplain-lacustrine facies. Rare channel sandstones that consist of about equal parts of sandstone, siltstone, and shale probably represent deposits in abandoned channels.

Thick sequences of alternating beds of siltstone, shale, carbonaceous shale, ironstone, and fine-grained sandstone commonly border channel sandstones, but also are repeated vertically and laterally in overlapping arrangements. The sandstone and siltstone show ripple drift and lenticular ripple laminations. This lithologic sequence commonly is rooted and contains tree stumps in growth position; such stumps were recognized by Allen (1965) as important features of a levee facies. The levee resulted from settling of suspended load during repeated overbank flooding. The interbedding of fine- and coarse-grained deposits reflects the fluctuating flood water stages, and the overlapping of these deposits suggests localized overtopping of low-lying areas of the levees.

Closely associated with the channel-levee facies is a minor amount of channel-plug and cutbank-slump facies. The channel-plug facies has a scoured base and consists of interbedded mudstone, siltstone, silty sandstone, and carbonaceous shale. It occurs either within a channel sandstone complex or as an isolated body, which usually lies within the floodplain facies. Both types of channel-plug facies represent infilling of abandoned channels mainly by suspended detrital and organic sediments washed in during floods. The detrital sediments contain ripple laminations, burrows, and root marks. The cutbank-slump facies is basally bounded by a curved fault surface. This facies consists of interbedded sandstone, siltstone, mudstone, carbonaceous shale, and coal that dip steeply (more than 30 degrees). The fault surface served as a rotational plane along which originally horizontal beds of floodplain and backswamp facies were tilted. The cutbank-slump facies resulted from the caving in of floodplain and backswamp facies as lateral erosion by the river undermined the bank at the water's edge. The channel-plug and cutbank-slump facies are related to highly sinuous streams, which mark the margins of the belt of channel-levee facies.

Floodplain Facies

The floodplain facies grades laterally into the channel-levee facies. The floodplain facies consists of interbedded sandstone, siltstone, mudstone, limestone, and ironstone. These sequences commonly contain freshwater mollusk fossils (Flores, 1981). The floodplain facies may be divided into crevasse-splay and lake-splay facies.

The crevasse-splay facies is either a single coarsening-upward sequence of shale, siltstone, and sandstone or a stack of coarsening-upward sequences. The sandstone occurs as a sheet-like deposit that can be divided into crevasse-channel and crevasse-splay types. The crevasse-channel sandstone, which is a fining-upward sequence, is basally erosional, is very fine to medium grained, and contains small-scale trough crossbeds and ripple laminations. The crevasse-channel sandstone probably represents secondary drainage conduits of the crevasse splay that served to transmit sediments to distal areas. The crevasse-splay sandstone is a very fine grained to medium grained, coarsening-upward deposit and has gradational or sharp basal contacts. Internal structures of crevasse-splay sandstones are dominated by various combinations of climbing ripple (ripple drift) lamination, asymmetrical ripple lamination, parallel lamination, small-scale cross lamination, and convolute lamination. A few degassing and pelecypod-escape structures are present, and smooth-walled vertical burrows and root marks are common. Rippled silty sandstones commonly are intercalated in the crevasse-splay sandstone. The abundant ripple laminations suggest deposition by slow-moving currents on margins of bodies of standing water, such as lakes. That the influx of detritus into the crevasse splay was episodic and triggered by flood stages is indicated by the silty intercalations and by the escape structures left where pelecypods sought to elude rapid burial.

The crevasse-splay sandstone is underlain by coarsening-upward shale and siltstone, interbedded with ironstone layers that contain abundant vertical and horizontal burrows. Shale and siltstone beds in this sequence are relatively thick near the proximal end of the crevasse splay and become thinner toward the distal end. The overall crevasse-splay deposit probably formed when floodwaters breached levees and deposited fan-shaped sediments onto the floodplain. The vertical stacking of crevasse-splay deposits resulted from overlapping progradation and lateral shifts of the splay lobes. Continuous progradation of the crevasse splays into distal parts of the floodplain permitted some of them to debouch into ephemeral lakes. This process of deposition results in the accumulation of lake-splay or crevasse-delta facies.

The lake-splay or crevasse-delta facies consists of sequences of shale, siltstone, and silty sandstone separated by limestones. These sequences contain freshwater mollusk fossils. The shale, siltstone, and silty sandstone are ripple laminated and burrowed. Some vertical burrows represent escape structures of pelecypods. Pulses of sediment that discharged into ephemeral lakes from crevasse splays during flood stages probably disturbed bottom fauna and resulted in their catastrophic burial or rapid escape. Generally, the limestone is mud-supported, thin, and lenticular and contains freshwater mollusk fossils. Limestone that has a high detrital content is commonly unfossiliferous. The limestone probably represents precipitation of carbonate zones in ephemeral lakes during cessation of detrital influx or nonflood condition, and/or in distal parts of these lakes not subjected to detrital sedimentation. Nonetheless, interruption by some detrital sediments in distal parts of the lakes is not common. Localized as well as widespread distribution of mollusk fossils occur in the floodplain facies. These fossiliferous zones, marking sites of lakes, suggest that lacustrine processes were active in the floodplain to various degrees, perhaps controlled by local or regional subsidence and compaction of underlying sediments. Biostratigraphic and biofabric studies of the freshwater mollusk fossils (John Hanley, oral commun.,) can document the evolutionary development of the lakes in terms of their depth, size, and relationship to sedimentation.

Backswamp Facies

The backswamp facies, which forms in the floodplain, is closely related to the floodplain facies. The major deposits of the backswamp facies are coal and carbonaceous shale, which consists of clay- and silt-size particles mixed with varying proportions of organic material. The coal beds, ranging from lignite to bituminous, underlie, overlie, grade laterally into, and interbed with carbonaceous shale beds. A minor amount of organic-rich siltstone and sandy siltstone forms partings in the coal beds. The coal beds associated with fluvial deposits of the intermontane basin fluvial system are usually thicker than those associated with deposits of the coastal plain-piedmont setting. Individual coal beds, which can be traced laterally along outcrop for a few miles before they split or merge, are also more laterally extensive in the intermontane basin fluvial deposits than in coastal plain-piedmont deposits. The discontinuity of coal beds associated with the backswamp facies resulted from their erosion by channels, splitting by overbank detrital sediments, and thinning over the channel-levee facies. The latter situation suggests that abandoned channel-levee facies served as topographically high alluvial ridges that frequently were encroached upon by adjoining backswamp.

Coal beds associated with the floodplain facies are as laterally continuous as those associated with the channel-levee facies but are prone to splitting by crevasse-splay facies. Thick coal beds occur at both the proximal and distal ends of the floodplain facies; however, the coals are thickest at locations distant from the channel-levee facies, far removed from detrital influx.

The most significant differences among coal beds in the intermontane basin fluvial depositional setting can be explained by the following interrelated factors:

1. Avulsion-controlled aggradation of the alluvial plain.
2. Prolonged vertical and lateral aggradation of fluvial channels.
3. Subsidence due to basement tectonic control.
4. Differential compaction of the various suites of fluvial facies.
5. Peat accumulation in raised (domed), ombrogenous swamps.
6. Length of time of peat accumulation.
7. Nature of the backswamp paleoflora.
8. Paleoclimate.

Perhaps the most important factors controlling the accumulation of anomalously thick coal in intermontane fluvial settings are the development of raised swamps and avulsion-controlled aggradation of the alluvial plain. An area of the alluvial plain that has been abandoned by the fluvial system is an ideal setting for establishment of poorly drained backswamps that are relatively free of detrital influx for a long period of time. Although this process is common to fluvial systems of both the coastal plain-piedmont and the intermontane alluvial plain, the effects are pronounced in the latter environment because of its inherent active tectonic setting. A clue to the origin of very thick Powder River Basin coal beds lies in their organic composition. That is, a forested backswamp may generate thick peat consisting chiefly of relatively incompactible woody organic debris. Today, densely forested backswamps are common in alluvial plains in Borneo, where they generate domed peat deposits perched above drainage levels. These raised or domed peat swamps, which rise as much as 15 ft above water levels of adjacent streams, are formed in high rainfall tropical regions. Here, the raised swamps are maintained by a near-surface ground-water table resulting from a high rate of rainfall (ombrogenous swamp). Similar conditions may have existed in the intermontane fluvial setting of the Powder River Basin, where coals have been determined to contain as much as 95 percent vitrinite or huminite (Peter D. Warwick, Univ. of Kentucky, oral commun.). Thus, it is not difficult to envisage the accumulation of very thick coals in the Powder River Basin because of their abundant woody composition.

COASTAL PLAIN-PIEDMONT FLUVIAL FACIES

The Rocky Mountain coal basins contain many Cretaceous and Tertiary fluvial coal-bearing deposits that were formed in a coastal plain setting bounded by the coastlines of epeiric and epicontinental seas on one side and Cordilleran uplift-piedmont areas on the other side. This coastal plain-piedmont setting was drained primarily by short-headed streams that were less than 200 mi long. The short-headed stream drainages characterized the coastal plain setting of the Cretaceous and Tertiary deposits in the Gallup sag and the Raton Basin. These fluvial settings differ from the coastal plain fluvial setting of Tertiary deposits landward of the epicontinental Cannonball sea (Paleocene Fort Union Formation) of the Williston Basin in North and South Dakota, which connected upstream to a intermontane trunk-tributary fluvial system in the Powder River Basin and western Montana.

Gallup Sag

The most important coal-bearing interval of the Cretaceous Crevasse Canyon and Menefee Formations forms the undifferentiated Gibson Coal Member of the Crevasse Canyon Formation and Cleary Coal Member of the Menefee Formation. These coal members contain economic coals whose occurrence is controlled by their environments of deposition (Cavaroc and Flores, 1984;

Flores, 1984). Within 25 mi, major coal depositional settings change from alluvial, through delta plain and open lagoon, into the reworked barrier complex of the Cretaceous Point Lookout Sandstone (Fig 2). These depositional changes occur southwest to northeast from the Mentmore area (southwest of Gallup), through the Gibson Canyon area (north of Gallup), to the Coal Mine Canyon area (northeast of Gallup), respectively.

Mentmore Area

The undifferentiated Gibson and Cleary Coal Members grade laterally in the Mentmore area into the Bartlett Barren Member of the Crevasse Canyon Formation (Fig. 3). The Bartlett Member is characterized by scarcity of coal and a high proportion of channel sandstone to fine-grained detritus.

The Bartlett Barren Member contains about 70 percent sandstone, the bulk of which is channel facies. This facies consists of channel sandstones that are mainly multiple stacked, are basally conglomeratic, and grade laterally and vertically into interbedded sandstone, siltstone, shale, carbonaceous shale, and a few thin coals of the floodplain and backswamp facies. The Bartlett Barren Member probably represents active, predominantly well-drained alluvial floodplain facies. The channel sandstones were not deposited in freely meandering channels but rather in channels that shifted by avulsion. Small channel sandstones represent transient crevasse-feeder channel deposits that originated from contemporaneous major channels. The fine-grained detritus juxtaposed with the channel sandstones represent well-drained levee and floodplain facies. Reducing environments within the alluvial floodplain appear to have continually received detritus from adjacent streams, resulting in the accumulation of carbonaceous shale and dirty coal beds.

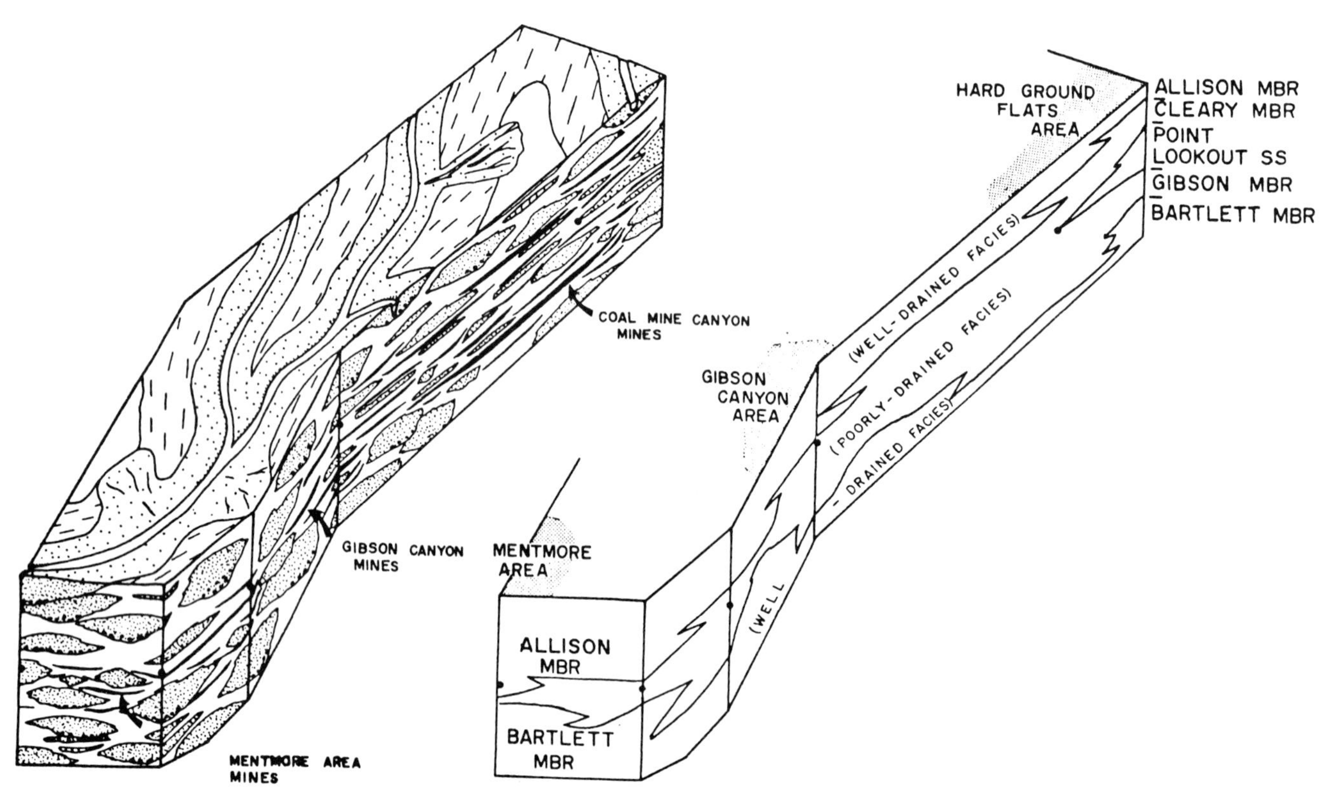

Figure 2. Environmental-stratigraphic framework of the Crevasse Canyon and Menefee Formations in the Gallup sag, New Mexico. From Cavaroc and Flores (1984).

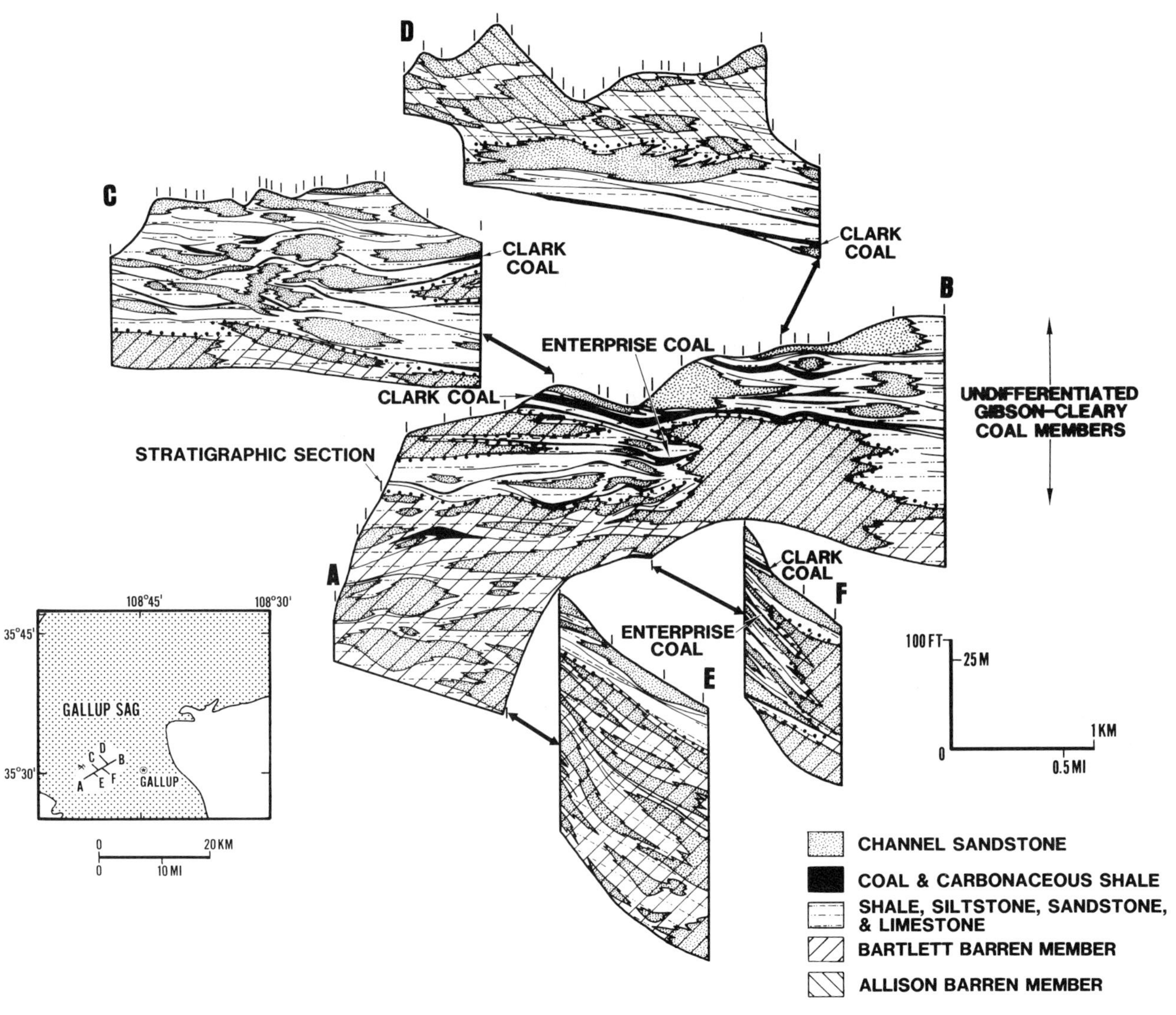

Figure 3. Network of stratigraphic sections of the Crevasse Canyon and Menefee Formations in Mentmore area, New Mexico. From Flores (1984).

The coal-bearing Gibson-Cleary interval in the Mentmore area, locally intertonguing with the Bartlett Barren Member, consists of interbedded channel sandstone and levee-floodplain facies of siltstone and shale as well as backswamp facies of carbonaceous shale and coal. Carbonaceous shale and coal beds commonly associated with the fine-grained deposits are as much as 7.5 ft thick and are laterally traceable in outcrops for as much as 2 mi. The channel sandstone maintains the same lenticular geometry and sequence of internal structures seen in sandstones of the Bartlett, but the sandstone is not as common as in the Bartlett. The depositional setting of the undifferentiated Gibson and Cleary Coal Members is interpreted as a part of the Bartlett alluvial plain abandoned by major channel activity. The waning of fluvial activity characterized by minimal detrital influx directly influenced the accumulation of thick peat in the poorly drained backswamps that followed shift and abandonment of channel activity.

Gibson Canyon Area

The undifferentiated Gibson and Cleary Coal Members in the Gibson Canyon area are underlain by the Bartlett Barren Member and, in general, have rock types similar to those in the Mentmore area (Fig. 2). However, their lithogenetic types are distinctly different. The crevasse-splay facies is more abundant in the Gibson Canyon area. This facies is also commonly overridden by small-channel facies. The second major difference is the presence of burrowed zones, a feature not found in the Mentmore area. Burrows are intense and are found both in clean sandstones interbedded with carbonaceous siltstones and in sandstones of the crevasse-splay facies. As at Mentmore, the lowest coal beds of the Gibson and Cleary Coal Members pass laterally into the Bartlett Barren Member. However, the major coal beds of the undifferentiated coal-bearing interval are thick, laterally extensive (on the order of several miles), and associated with widespread carbonaceous shale zones.

The thick, laterally continuous coals of the undifferentiated Gibson and Cleary Coal Members in the Gibson Canyon area are interpreted as accumulations over embayed, drowned surfaces of earlier alluvial deposits. The bifurcation of the primary fluvial channels into distributaries suggests that the embayed areas represent interdistributaries on a upper delta plain. The intensely bioturbated sequences associated with these embayed areas indicate that these bays were built into poorly drained swamp platforms.

<u>Coal Mine Canyon Area</u>

In the Coal Mine Canyon area, the Gibson and Cleary Coal Members are split by the landward pinchout of the Point Lookout Sandstone (Fig. 2). Deposits of the Cleary Coal Member, which overlie the Point Lookout Sandstone, consist of thin, rippled, burrowed sheet sandstones associated with carbonaceous shales. These deposits give way to small, lenticular channel facies of sandstones associated with coarsening-upward detritus. Coals associated with these deposits are thin and discontinuous. The coal beds are thicker and are laterally more continuous in the upper part of the interval.

The Gibson Coal Member, beneath the Point Lookout Sandstone, contains sheet-like rippled and burrowed sandstones that are interbedded with small, lenticular channel sandstone bodies. Carbonaceous shales contain intense bioturbation and are interbedded with clean silty sandstone and carbonaceous siltstone. Coal beds are thick and are laterally discontinuous. The coal beds in the lower part of the Gibson Coal Member pinch out and become discontinuous into the Bartlett Barren Member.

The Point Lookout Sandstone in the Coal Mine Canyon area was a physical barrier that isolated carbonaceous lagoonal deposits of the Gibson Coal Member, which in turn formed across abandoned fluvial facies of the Bartlett Barren Member. The barrier sandstone complex was oriented from northwest to southeast. The abundant southwesterly dips of crossbeds and the intense bioturbation of the tops of sandstone beds in the Point Lookout indicate deposition in back-barrier washover and/or tidal delta environments. Lagoonal conditions at the back of the barrier complex were subsequently destroyed by the northerly influx of detritus beginning in Cleary time. Reintroduction of deltaic and alluvial activities seaward behind the regressive Point Lookout Sandstone permitted optimum conditions for the accumulation of thick coal in late Cleary time.

Interpretations

Figure 4 shows successive paleogeographic reconstructions of (1) temporally equivalent lithofacies of the intertongued Bartlett Barren Member and undifferentiated Gibson and Cleary Coal Members and (2) the Clark coal bed and laterally equivalent rocks in the Mentmore area (Cavaroc and Flores, 1984; Flores, 1984). In the intertongued interval, discontinuous, poorly drained backswamps are disrupted by random crevassing from meandering channels of the lower alluvial plain. The influx of detritus and a lowering of the ground-water table between river flood stages curtailed the accumulation of thick peat deposits in the backswamps. As the deposits of the intertongued interval were abandoned during a westward shift of channel activity, interfluvial poorly drained backswamps developed and encroached from the east-northeast, forming the thick, laterally extensive coal beds in the lower alluvial plain. The poorly drained backswamps are part of the western reaches of broad backswamps that formed between northeast-trending

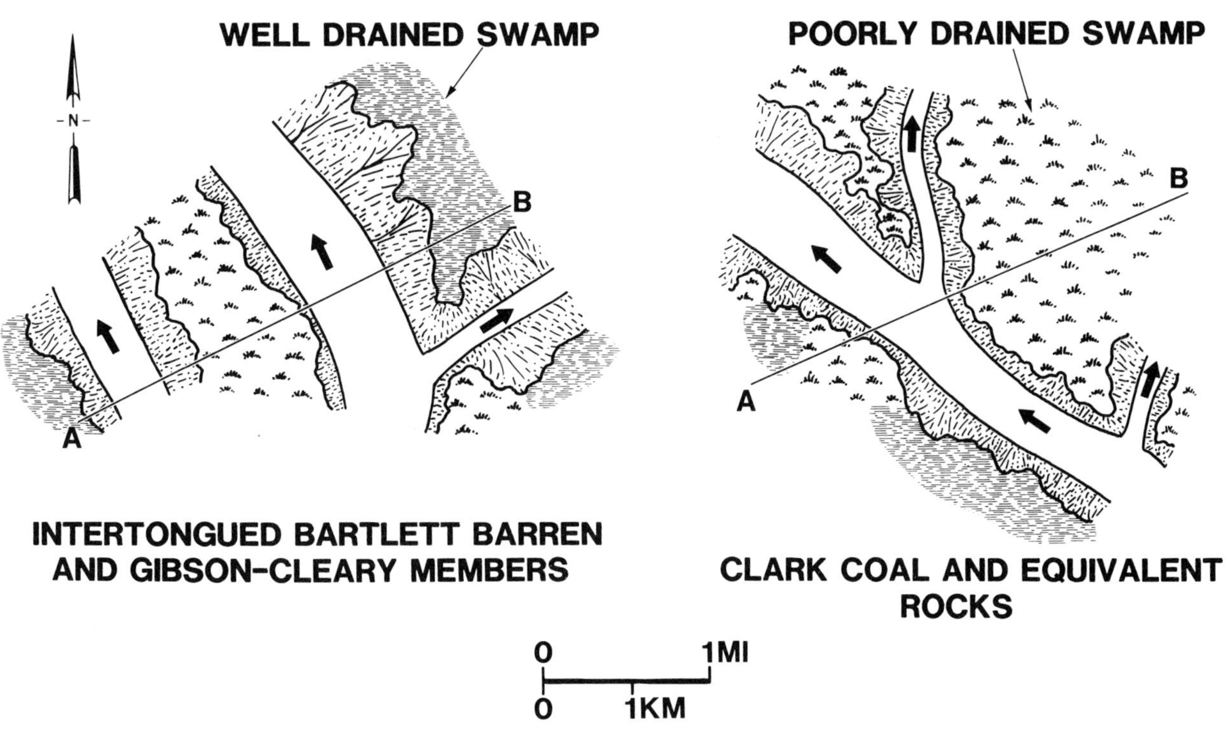

Figure 4. Paleogeographic reconstructions of intertongued Bartlett Barren, Gibson, and Cleary Members and Clark coal and equivalent rocks. From Flores (1984).

channels. A regional analysis of paleogeography (Flores, 1984) indicates that the westward development of poorly drained backswamps in a lower alluvial plain environment corresponds to back-barrier lagoonal transgression across embayed areas and to the maximum landward advance of barrier islands. Eastward or seaward, the lower alluvial plain grades into deltaic and barrier complexes. Lagoons, which are seaward of deltas and landward of barrier islands, encroached over deltaic deposits as far as 6 mi east of the lower alluvial plain. Thus, the trangression of lagoons may have transformed lower alluvial plain backswamps into poorly drained environments and sustained their development for a prolonged period.

The succession of paleogeographic reconstructions for the lower alluvial plain and the adjoining coastal area is shown in Figure 5. The paleogeographic reconstructions illustrate random avulsion or autocyclic processes in the lower alluvial plain, where channels were successively diverted from courses subparallel to the coastline into courses perpendicular to the coastline. These models, which show widespread development of poorly drained backswamps over a well-drained alluvial plain, suggest a regional rise of the ground-water table that sustained the swamps. A rising ground-water table coincided with the lagoonal transgression that accompanied the maximum landward advance of deposits of the Point Lookout barrier complex. That this rise of the ground-water table was maintained for a long time is suggested by stacking of barrier deposits along the coast.

Raton Basin

Coal deposition in an upper alluvial plain of a coastal plain-piedmont setting, such as the Raton Basin, was different from that in a lower alluvial plain near a maximum transgressive barrier coastline, such as the

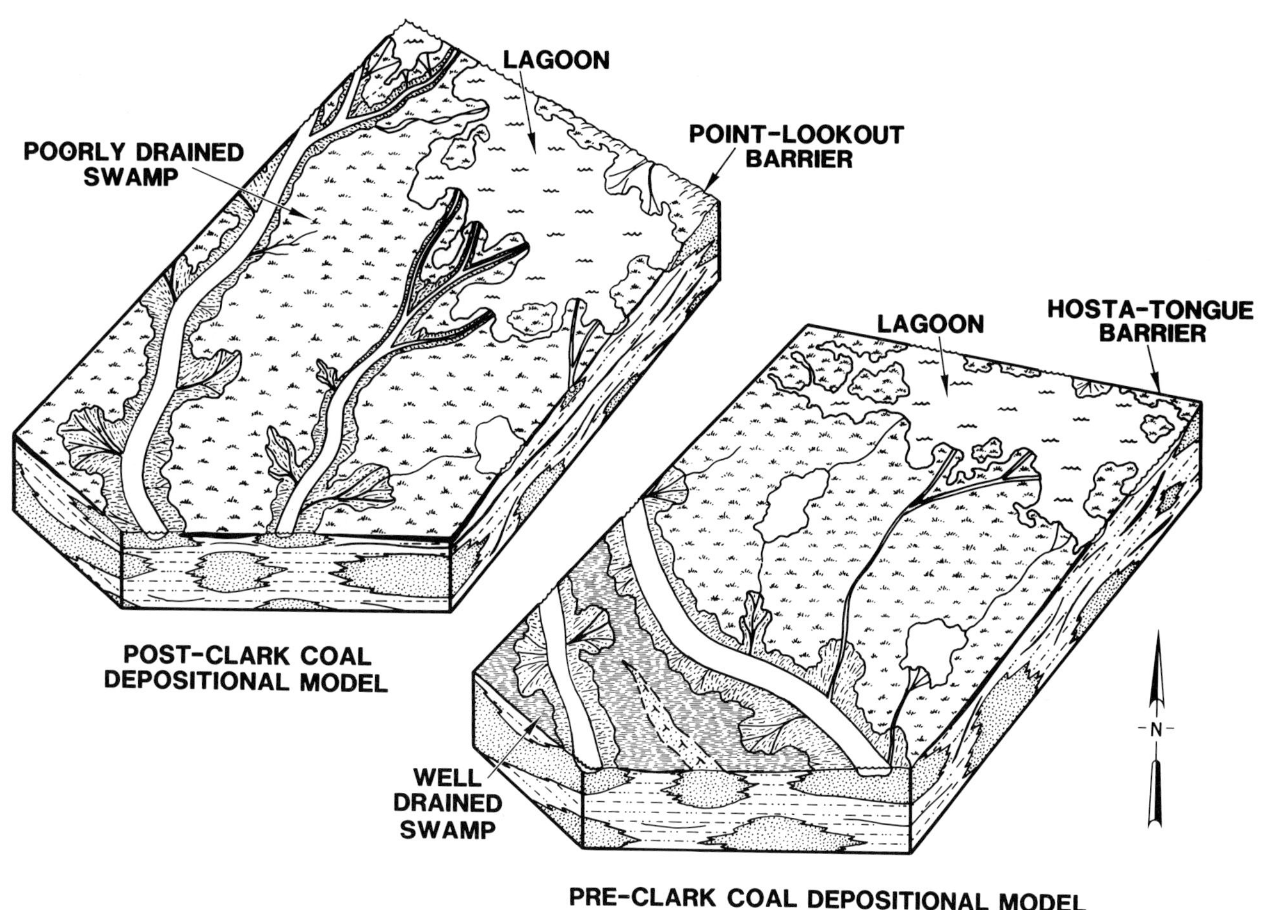

Figure 5. Depositional models of the lower alluvial plain and coastal area in the Gallup sag. From Flores (1984).

Gallup sag. The fluvial coal deposits in the Cretaceous and Tertiary rocks in the Raton Basin are mainly contained in the Raton Formation. The Raton Formation is overlain by and merges with the Poison Canyon Formation, which is barren of coal beds. Poison Canyon is composed mainly of conglomeratic and coarse-grained sandstone of the channel facies and of reddish-green mudstone, siltstone, and silty sandstone of the levee and floodplain facies. The Raton Formation is underlain by the Vermejo Formation, which consists of coal-bearing deltaic and back-barrier deposits and in turn is underlain by delta- front and barrier deposits of the Trinidad Sandstone and prodelta and shelf deposits of the Pierre Shale. The Raton Formation can be divided into a lower coal zone, a barren series, and an upper coal zone (Pillmore and Flores, 1984).

Lower Coal Zone

The lower coal zone contains most of the Cretaceous part of the Raton Formation, inasmuch as the Cretaceous-Tertiary boundary is at or near the top of the coal zone (Pillmore and Flores, 1984). The lower part of this zone consists of a prominent ledge-forming channel facies that includes conglomeratic and quartzose sandstone components (Fig. 6). This channel facies grades westward into conglomeratic beds that are not differentiated from the Poison Canyon Formation. The remainder of the lower coal zone is composed of fine-grained channel-levee facies interbedded with floodplain facies of silty sandstone, siltstone, and mudstone. The backswamp facies includes thin to thick beds of carbonaceous shale and coal.

The occurrence of the conglomeratic and quartzose channel facies suggests tectonic control of fluvial sedimentation in the upper alluvial plain. The conglomeratic channel facies probably represents deposition in basin-margin, high-bedload braided streams that merged basinward into meandering streams. The quartzose channel facies, which contains as much as

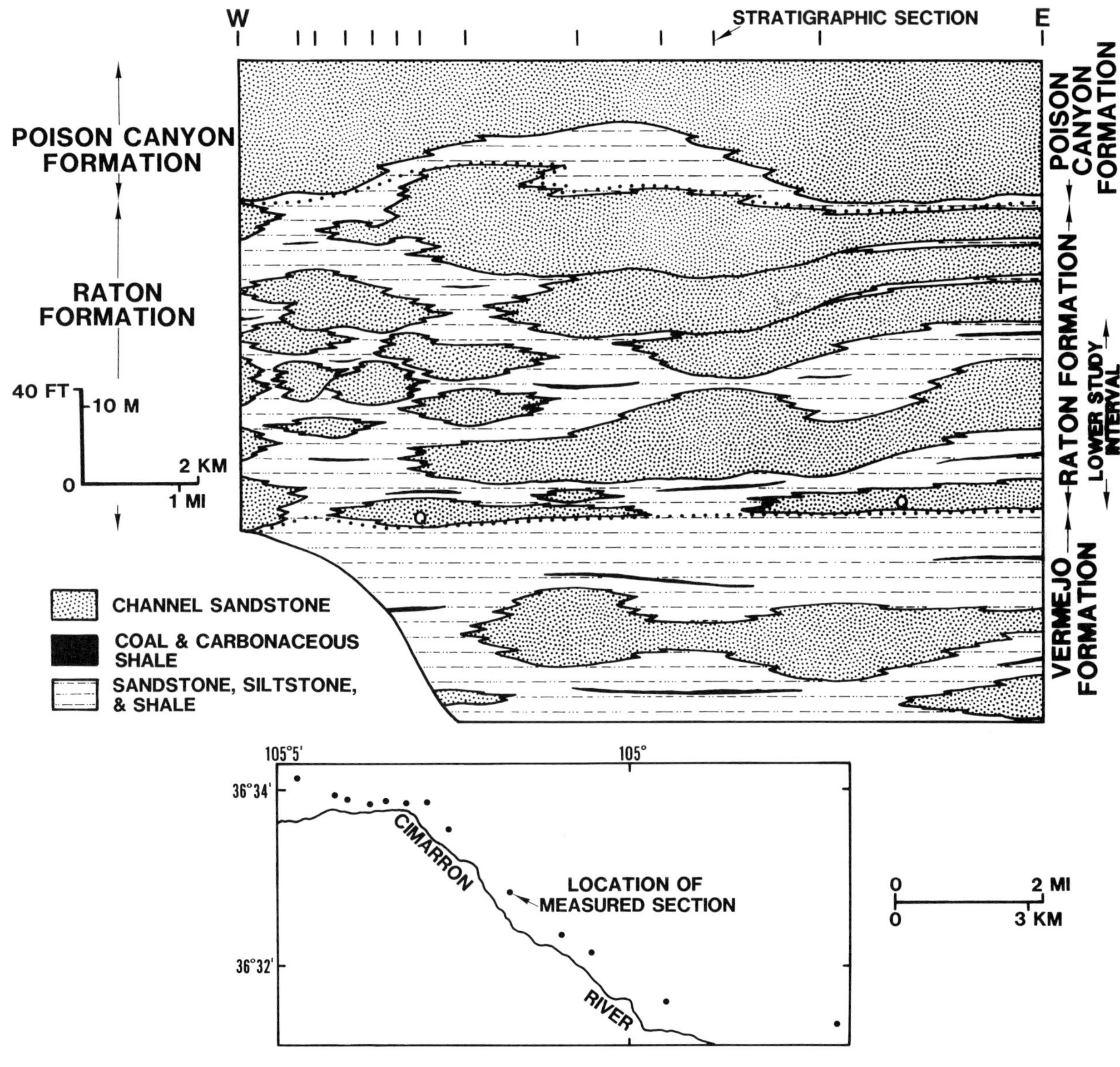

Figure 6. Stratigraphic cross section showing lithofacies of the Raton and Poison Canyon Formations; quartzose (Q) channel sandstones are shown in lower part of lower coal zone. From Flores (1984).

98 percent quartz, suggests recycling of the Lower Cretaceous, quartzitic Dakota Sandstone (Flores and Tur, 1982). The influx of conglomeratic and quartzose sediments into the fluvial system indicates uplift and erosion of the source area. This process created basin-margin braided streams and choked the floodplain with overbank sediments, thereby affecting the accumulation of peat in associated backswamps.

Barren Series

The barren series is dominated by a channel facies, which thickens and coarsens into conglomeratic units toward the west, where it merges with the Poison Canyon Formation (Fig. 6). The channel facies varies from clustered, offset channel sandstones at the east (Fig. 7) to sheetlike, conglomeratic channel sandstones at the west (Fig. 6). However, these channel-facies types are locally coeval in parts of the basin. The barren series is characterized by sparse coal and carbonaceous shale beds. These organic deposits, which are thin and discontinuous, are interbedded with minor deposits of silty sandstone, siltstone, and mudstone, as well as with the channel sandstone.

The barren series of the Raton Formation is interpreted as deposits of high-bedload meandering streams (e.g., offset channel sandstone) and braided streams (e.g., sheetlike conglomeratic channel sandstone). Where these deposits merge with conglomeratic beds of the Poison Canyon Formation, the braided stream deposits pass into alluvial fan deposits. The conglomeratic beds, particularly the conglomeratic and quartzose sediments of the lower coal zone, resulted from uplift and erosion of the source area. However, unlike the pulse of coarse detritus in the lower coal zone, the barren series represents a more extensive influx of conglomeratic detritus. The subordinate floodplain and backswamp facies of the barren series may reflect rapid lateral migration of stream channels that reworked the facies.

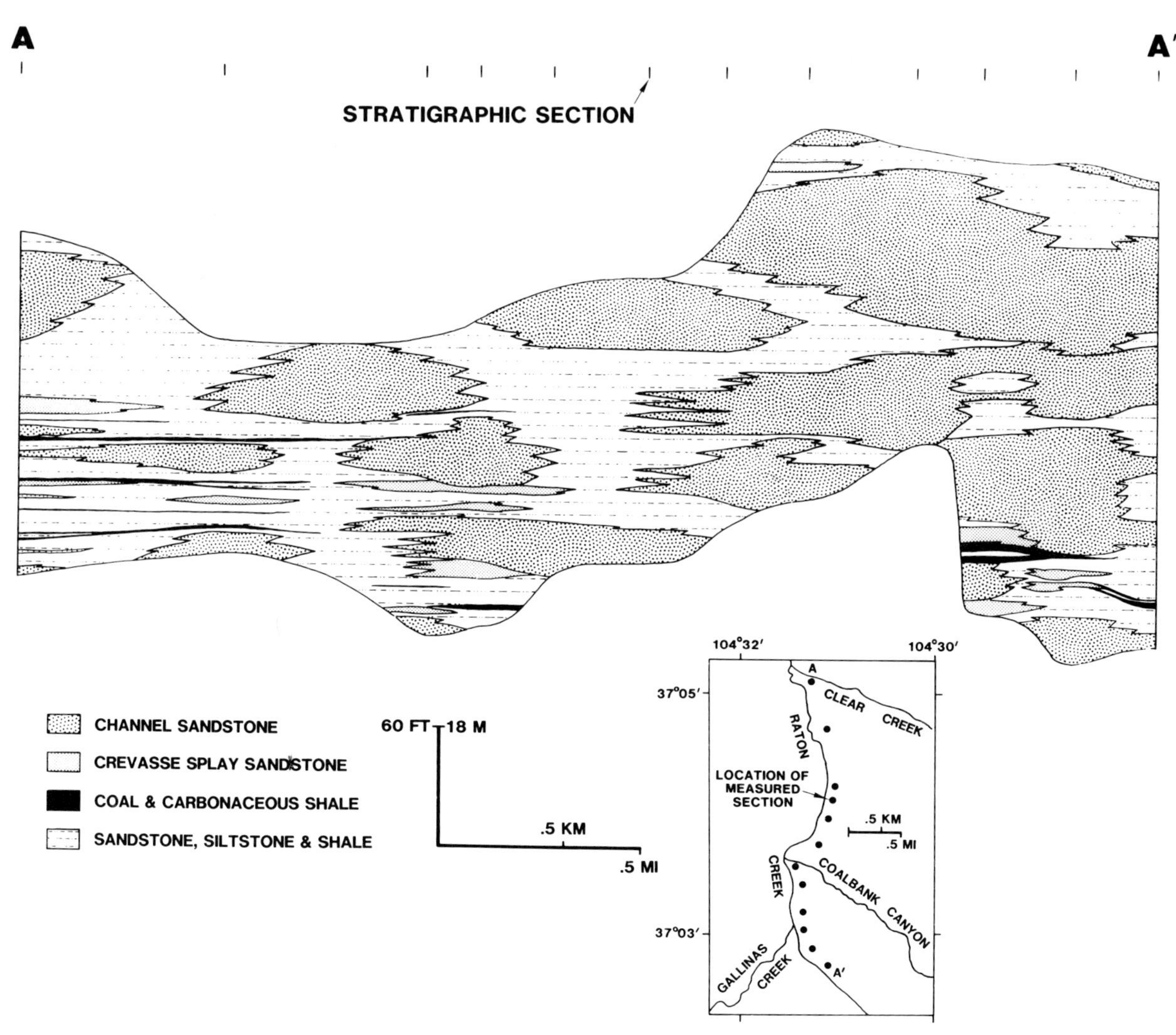

Figure 7. Stratigraphic cross section showing lithofacies of barren series of the Raton Formation. From Flores (1984).

Upper Coal Zone

The upper coal zone includes abundant backswamp facies of carbonaceous shale and coal beds that are thick and laterally continuous (Fig. 8). This facies is commonly interbedded with thick floodplain facies of silty sandstone, siltstone, and mudstone. The bulk of the floodplain facies is composed of crevasse-splay deposits. The floodplain and backswamp facies grade into a channel-levee facies that is composed of single or multistory bodies. Channel-plug and cutbank-slump facies are commonly associated with the channel-levee facies. These facies mark the margins of the belt of channel-levee facies.

The facies of the upper coal zone represent deposits of meandering streams bounded by extensive floodplains and isolated poorly drained backswamps. The thick accumulation of floodplain facies reflects rapid aggradation resulting from active subsidence (tectonic and/or sediment loading), which permitted their preservation below the base level of erosion. The backswamps probably formed in distal parts of floodbasins, away from detrital influx, or on abandoned courses of meandering streams.

Interpretations

The architectural framework (vertical and lateral) of the upper alluvial plain deposits of the Raton Formation in the Raton Basin are controlled by the interaction of extrabasinal and intrabasinal tectonic settings (allocyclic processes). The interaction is displayed by vertical succession (Fig. 9) of the alternating sand-rich cycles (e.g., conglomeratic channel facies) and mud-rich cycles (e.g., floodplain-backswamp facies). The sand-rich cycles are interpreted as synorogenic deposits controlled by maximum effects of extrabasinal tectonism (e.g., uplift of source area) and minimum effects of intrabasinal tectonism (e.g.,

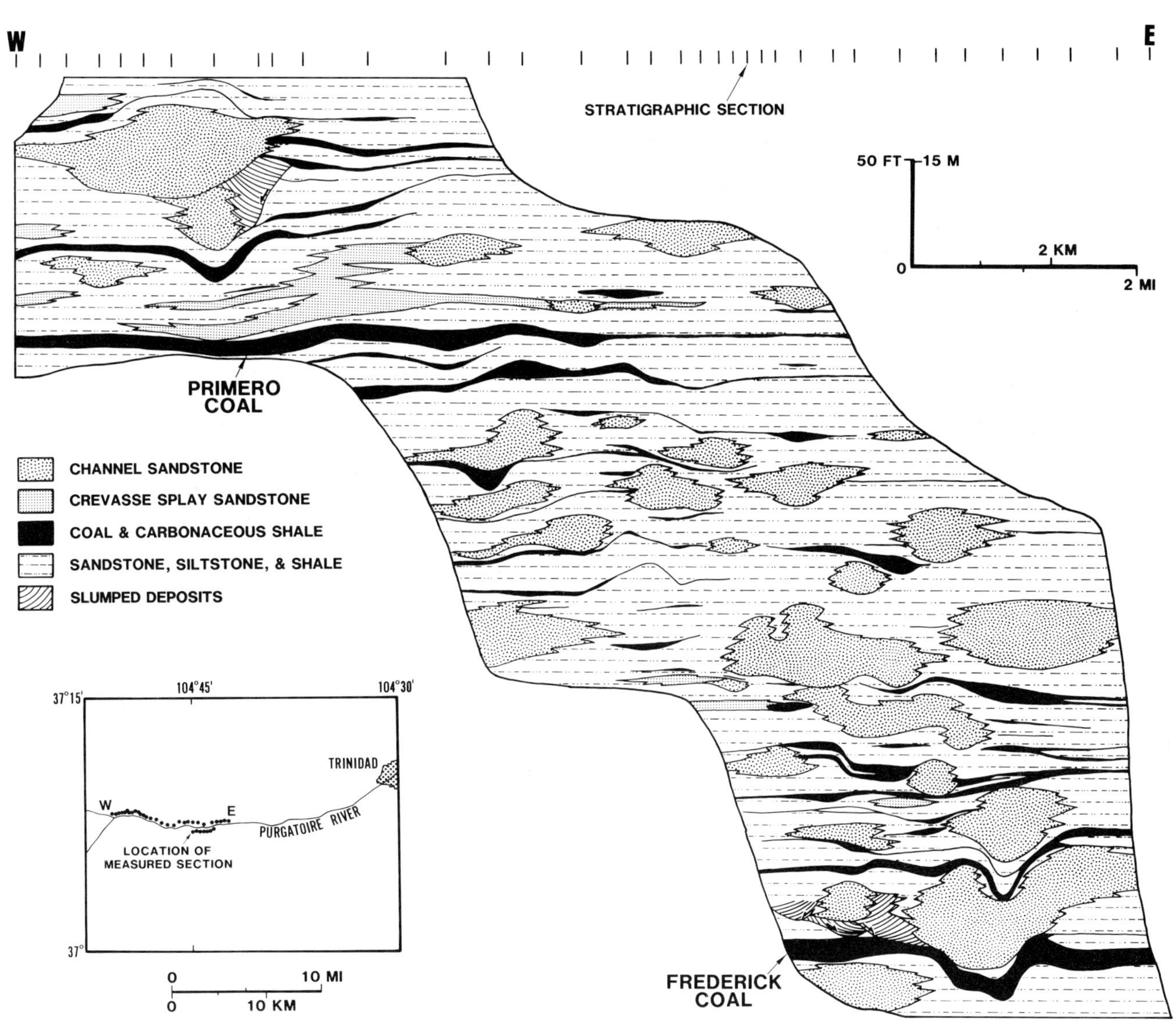

Figure 8. Stratigraphic cross section showing lithofacies of upper coal zone of the Raton Formation. From Flores (1984).

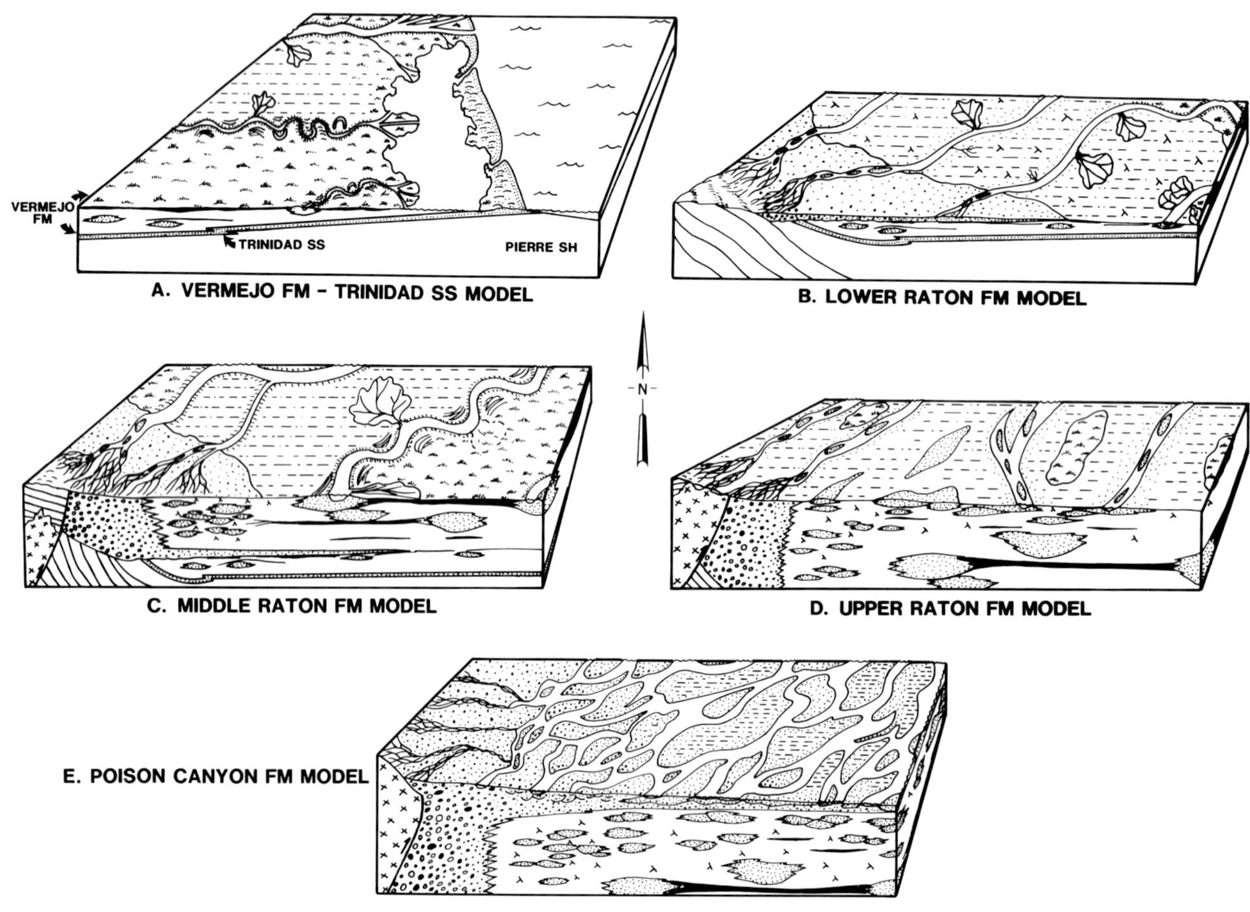

Figure 9. Block diagrams showing the influence of tectonism on sedimentation and environments of deposition from Pierre to Poison Canyon times.

subsidence). The coarsest clasts in the Raton Formation are found in this cycle; these reflect periods of aggradation by high-gradient streams. Rapid lateral migrations of these streams caused extensive reworking of the floodplain and backswamp facies permitting only minimal preservation of these facies. The sand-rich cycle is characterized by progressive introduction of coarse clastics into the upper alluvial plain by alluvial fans, braided streams, high-bedload meandering streams. The uplift of the source area to the west (San Luis highland) may have been related to recurring pulses of thrust movements (Woodward and Snyder, 1976). Thus, the sand-rich wedges mark the sporadic orogeny and movement along thrust faults. As these orogenic movements diminished with time, the sand-rich cycles were replaced by mud-rich cycles.

Vertical aggradation in the upper alluvial plain occurred during deposition of the mud-rich cycles of the Raton Formation. These cycles, which contain a higher proportion of floodplain and backswamp sediments than the sand-rich cycles, represent deposition during periods of extrabasinal tectonic quiescence, probably between thrust movements. Active basin subsidence, probably due to sediment loading, developed at this time and promoted preservation of floodplain deposits. Depositional loading also may have forced ground-water out of areas of high pore pressures, such as the bottom of the basin, toward the top, or from areas of maximum deposition to basin margins (Kreitler, 1979). This dynamically compacting basin developed a hydrologic setting that sustained poorly drained backswamps in the upper alluvial plain.

INTERMONTANE FLUVIAL FACIES

Powder River Basin

The basic stratigraphy of the Eocene Wasatch Formation in the Powder River Basin consists of conglomeratic facies at the western basin margin, which grades into coal-bearing facies to the east. The conglomeratic facies includes the Kingsbury and Moncrief Members. These members consist of conglomerate interbedded with sandstone, siltstone, mudstone, and coaly carbonaceous shale. The Kingsbury Member contains more sedimentary rock fragments (Mesozoic and Paleozoic) than the Moncrief Member, which contains abundant igneous and metamorphic rock fragments (Precambrian). The variations in clast types of these conglomeratic facies are governed by the uplift and deep denudation of the source area. The conglomeratic facies passes basinward into interbedded coal, carbonaceous shale, mudstone, siltstone, limestone, and sandstone. The coal bed, which is a minor part of the Wasatch, is as much as 250 ft thick.

Conglomeratic Facies

The vertical and lateral variations in lithic types of the conglomeratic facies, typified by the Kingsbury Conglomerate Member, are shown in Figure 10. Measured sections A through E are described from the conglomerate-dominated part of the facies at the west to the sandstone-dominated part to the east. The conglomerate beds are scour based, imbricated, and graded from boulder to pebble size. These beds are interbedded with basally erosional conglomeratic sandstone that is trough and planar crossbedded. A few beds of sandy, mottled siltstone are interbedded with these conglomeratic units. Eastward or basinward, the conglomeratic units pass into interbedded conglomeratic sandstone, sandstone, and siltstone. The sandstone is basally erosional, trough and planar cross- bedded, and convolute laminated. The siltstone is variegated and root mottled. Farther basinward, siltstone interbedded with coaly carbonaceous shale becomes very common. Conglomeratic sandstone is replaced by a finer grained sandstone; a few beds of this sandstone are sheetlike and ripple laminated.

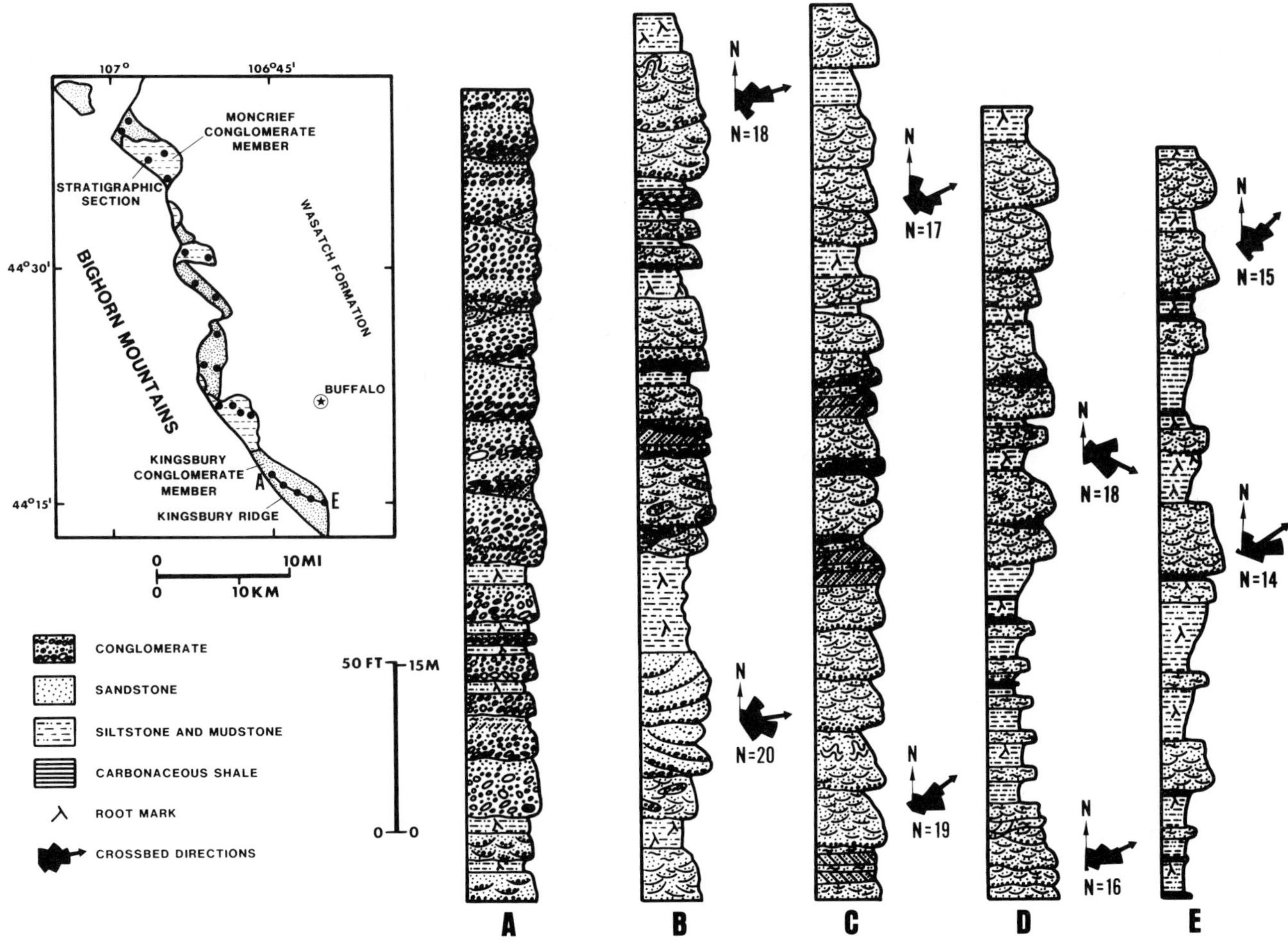

Figure 10. Lithofacies of the conglomeratic Kingsbury Member of the Wasatch Formation. From Flores and Warwick (1984).

The conglomeratic facies can be subdivided into proximal fan, midfan, and distal fan facies based on the relative proportion of conglomerate and the abundant finer grained detritus. The proximal fan facies is represented by the conglomerate-dominated sequence at the west, and the distal fan facies consists of the siltstone- and sandstone-dominated interval at the east. The sequence of mixed sandstone and conglomeratic sandstone between these extremes represents the midfan facies. The alluvial-fan facies displays downslope deposition of sheet-gravel bars in the proximal fan, braided channel sands in the midfan, and discrete channel sands separated by floodplain sediments in the distal fan. The paleoflow direction of the alluvial fans is primarily to the east and northeast, as indicated by trough crossbed measurements (Fig. 10).

Coal-Bearing Facies

The facies sequences and associations of the coal- bearing facies are shown in Figures 11 and 12. Figure 11 shows facies variations of the Felix coal zone and associated sediments, which are laterally equivalent to the lower part of the Kingsbury Member. Figure 12 shows facies characteristics of the Lake de Smet coal zone and associated sediments, which are equivalent to the lower part of the Moncrief Member. Both coal-bearing facies consist of channel-levee, floodplain, and backswamp facies; however, their facies associations differ at various locations in the alluvial plain. The Felix coal zone and associated deposits formed in the central part of the alluvial plain. The Lake de Smet coal zone and associated sediments were deposited at the margin of the alluvial plain near the alluvial fans.

The facies relationship (Fig. 11) of the Felix coal zone and associated sediments shows that the channel-levee facies is arranged en

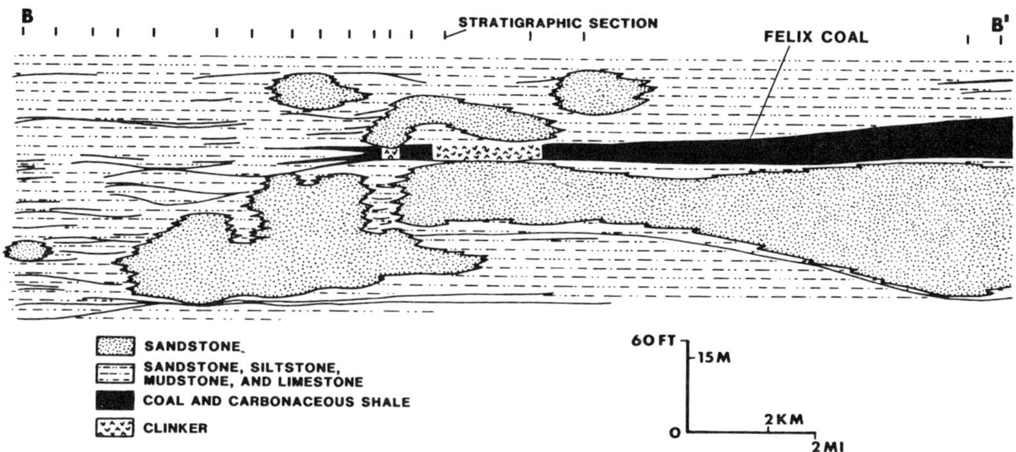

Figure 11. Stratigraphic cross section showing lithofacies of the Felix coal zone and associated sediments. From Flores and Warwick (1984).

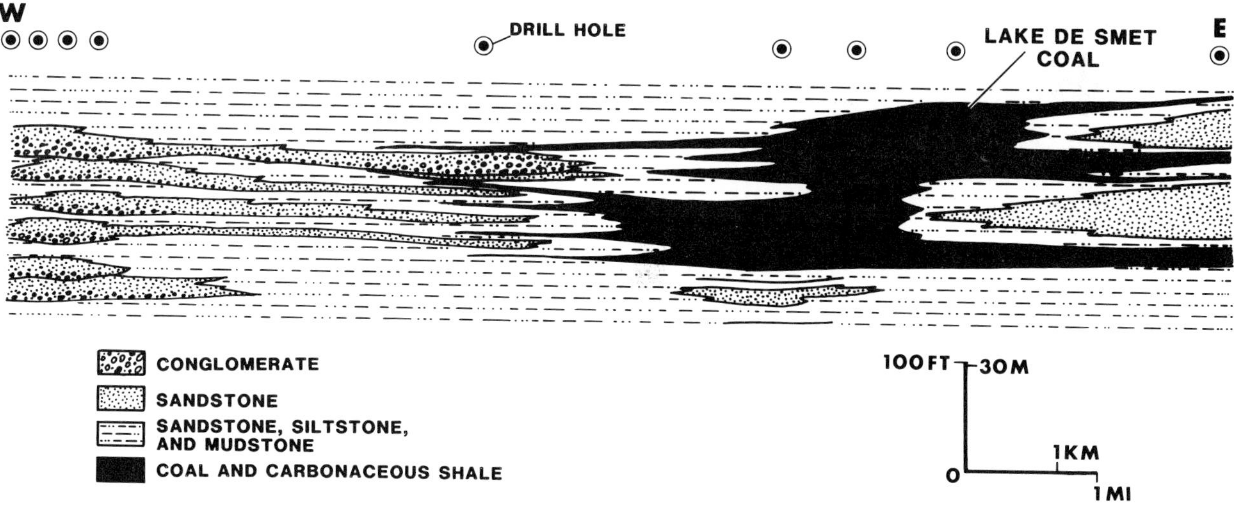

Figure 12. Stratigraphic cross section showing lithofacies of the Lake de Smet coal zone and associated sediments. From Flores and Warwick (1984).

echelon, forming a laterally extensive complex. This facies association is exemplified by the complex of channel-levee facies below the Felix coal zone. The coal zone is thick immediately above the channel-levee facies complex, and it thins and splits into several coal beds where the channel-levee facies complex grades into floodplain and backswamp facies. The channel-levee facies complex represents deposits of meandering streams which were subsequently abandoned because of avulsion. The abandoned meanderbelt deposits, which contain poorly compactible channel sands, remained higher than adjacent areas that were underlain by readily compactible fine-grained floodplain facies. The abandoned channel-levee facies complex served as a topographic high on which backswamps developed and peat accumulation was maximal; by contrast, in the low-lying backswamps, peat accumulation was minimal because of detrital influx. This alluvial-plain complex is about 25 mi east of the Kingsbury alluvial fans, which drained from the western highlands (ancestral Bighorn Mountains).

The facies association of alluvial fans with the alluvial plain is best exhibited by the Lake de Smet coal zone and associated sediments. The channel-levee facies complex, which probably formed in a meandering stream, splits the Lake de Smet coal zone into several coal beds to the east. Towards the west, the Lake de Smet coal zone is also split into several coal beds by the distal alluvial fan facies of the Moncrief Member. The Lake de Smet coal zone, about 2 mi wide and 17 mi long, is areally distributed between the channel-levee and distal fan facies. The accumulation of the Lake de Smet coal zone, which is as much as 250 ft thick, is an enigma. The location of the Lake de Smet backswamp, immediately between areas of active detrital sedimentation, should have served as a detriment to organic accumulation because of detrital overflow from these areas. Detrital influxes into the backswamp may, however, have been deterred by the development of raised swamps elevated above drainage levels. This condition and the inherently high permeability and transmissivity of the coarse-grained deposits of the adjoining channel-levee and distal-fan facies, which served as zones of groundwater discharge, very likely provided the ideal setting for accumulation of thick peat.

Interpretations

During the deposition of the Wasatch Formation in the Powder River Basin, sedimentation in the intermontane alluvial plain was controlled by tectonism. Uplifts of the source area to the west in the ancestral Bighorn Mountains along thrust faults (Blackstone, 1981) promoted deep erosion and produced sediments that were transported, via alluvial fans along the basin margins, to the alluvial plain in the basin proper (Fig. 13). Various environments in the alluvial plain, from abandoned, topographically high meanderbelts to floodbasins, formed poorly drained backswamps in which coals accumulated. In both styles of coal deposition, the absence of detrital influx into the backswamps was a major controlling factor. This condition was attained by the development of backswamps either on a depositional topographic high, such as poorly compactible channel sands, or on a raised platform well above adjacent drainage systems.

Finally, the paleohydrology of shallow- and deep-basin ground waters played an important role in the development of backswamps in the intermontane alluvial plain. Alluvial fans were recharge areas for shallow ground water that discharged into the alluvial plain. Like the Raton Basin, the Powder River Basin was probably formed as a dynamic compacting basin and, under this condition, deep-basin ground water flowed from the

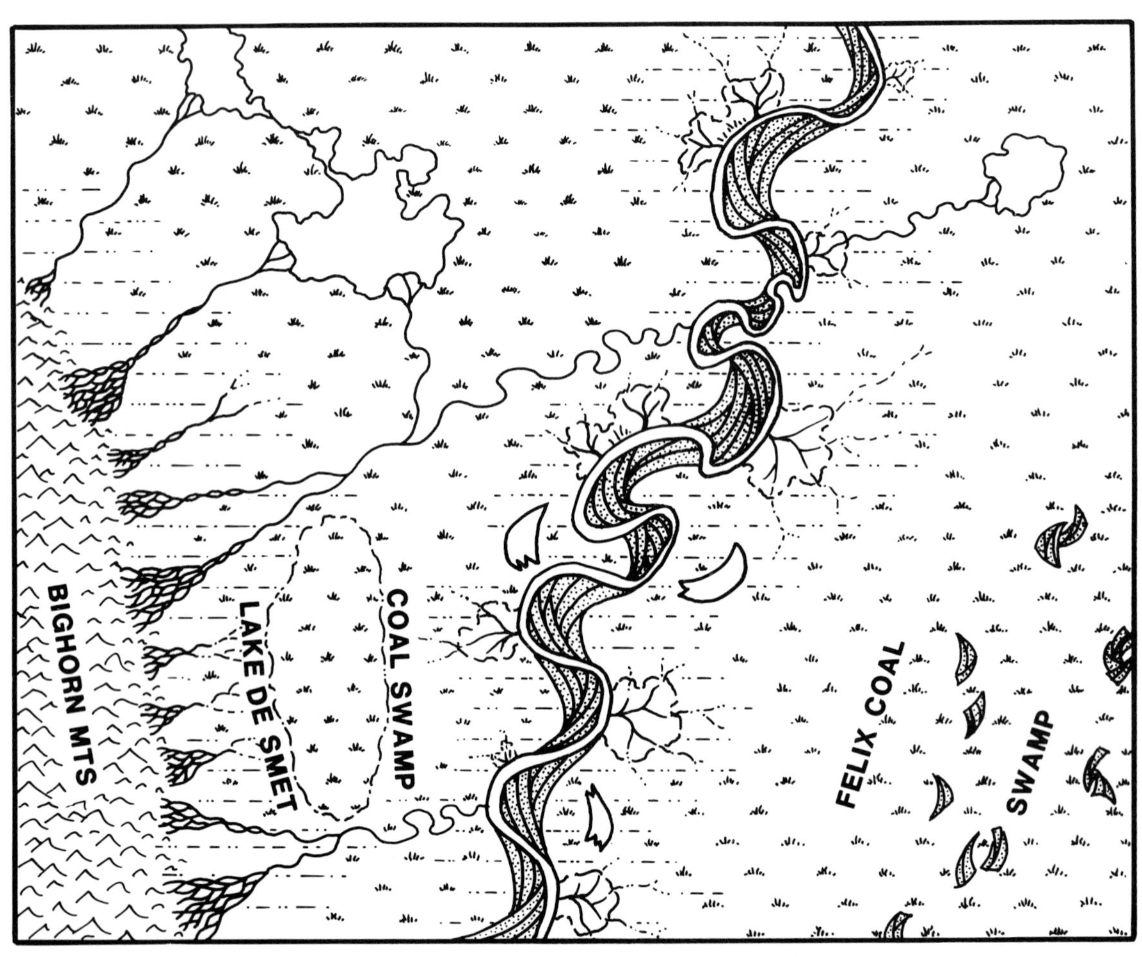

Figure 13. Reconstruction of depositional environments showing placement of the Felix and Lake de Smet coal swamps in an intermontane alluvial plain-alluvial fan setting. From Flores and Warwick (1984).

bottom to the top of the basin or from areas of maximum sedimentation to basin margins. These hydrologic conditions may have assisted in sustaining poorly drained backswamps of the intermontane alluvial plain.

ECONOMIC APPLICATIONS

Although the fluvial facies present in each of the study basins of the Rocky Mountain region vary greatly, the depositional architecture of framework channel-levee facies and nonframework floodplain-backswamp facies is very similar. That is, within an alluvial plain the framework channel-levee facies defines the axis of sand-rich sedimentation. This facies may occur as isolated single body, as offset nested bodies, and as sheetlike bodies. Within a fluvial axis, the framework channel-levee facies may form belts from a few hundred feet to several miles wide that are oriented along depositional dip. Thus, the sand-rich fluvial axes form excellent hydrocarbon reservoir bodies.

As demonstrated by the fluvial facies of the Gallup sag, the Raton Basin, and the Powder River Basin, the channel-levee facies is temporal to floodplain and backswamp facies. These facies are arranged either vertically, randomly, or en echelon, and their arrangements are controlled by autocyclic and allocyclic processes. The floodplain and backswamp facies are characterized by mud- and organic-rich deposits, which serve as seal and source facies. The organic deposits in the form of woody and herbaceous plant debris, and layered coal beds are potential biogenic and thermal gas-prone source facies. The proximity of the lower alluvial plain to the marine environment may lead to juxtaposition of the fluvial channel-levee facies to oil-prone source facies. The floodplain facies consists of crevasse-splay and channel sandstones encased in mud, which may also develop into excellent secondary reservoir bodies.

One of the most important aspects of viewing coal beds in an exploration context is their thickness, continuity, and areal distribution. These parameters assist in determining environments of deposition as well as facies distributions, especially in relation to framework channel-levee reservoir facies in coal-bearing deposits. In the Gallup sag, backswamps of the lower alluvial plain are readily distinguishable from those of the delta plain- lagoonal setting based on differences in the thickness, continuity, and areal distribution of the coal beds. Figure 14 shows outlines of underground workings of deep mines of the undifferentiated Gibson and Cleary Coal Members in the Gibson Canyon area. The boundaries of each deep mine define the limit outside of which the coal beds are not economical due to thinning, splitting, or absence of coal. Thus, the outlines of these mines depict small and large areal distributions of minable coal bodies. A small areal distribution (A and B) characterizes coal beds formed in backswamps of the lower alluvial plain. A large areal distribution (C and D) typifies coal beds that accumulated in backswamps of the deltaic-lagoonal environment. The limited areal distribution and rapid variation and discontinuity of coal beds of the lower alluvial plain resulted from random avulsion of crevasse splays into low-lying backswamps, which were transformed into major fluvial axes. The extensive areal distribution and less rapid change in thickness and continuity of coal beds of the deltaic-lagoonal environment resulted from formation on abandoned deltaic platforms transgressed by lagoons deep enough to develop peat swamps. The areal distribution of coals in the lower alluvial plain is exemplified by the Enterprise coal in the Mentmore area. The isopach map of the Enterprise coal (Fig. 15) displays a small areal distribution restricted by adjoining, laterally equivalent channel-

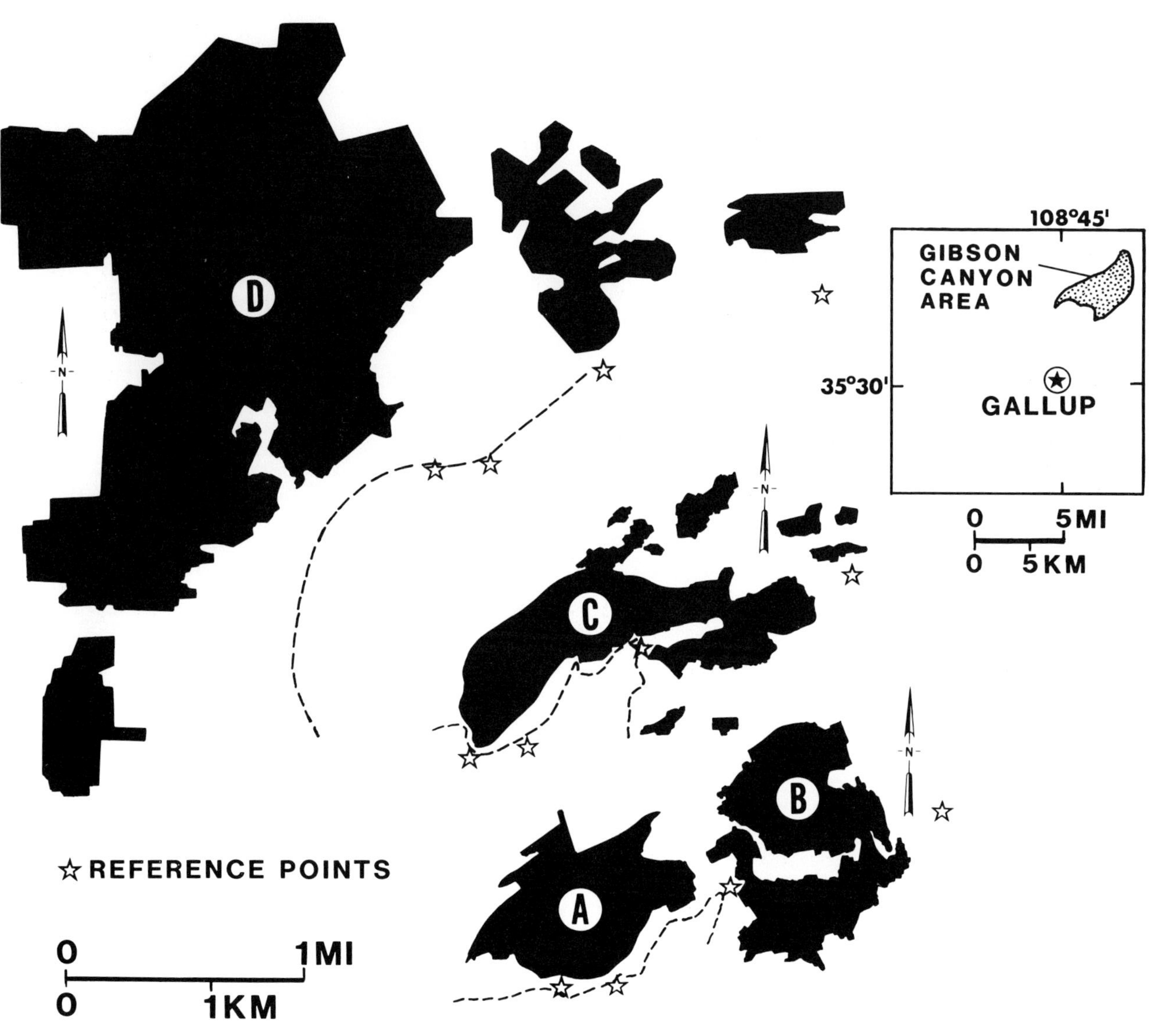

Figure 14. Outlines of underground workings of coal mines of the undifferentiated Gibson and Cleary Coal Members in the Gibson Canyon area. From Cavaroc and Flores (1984).

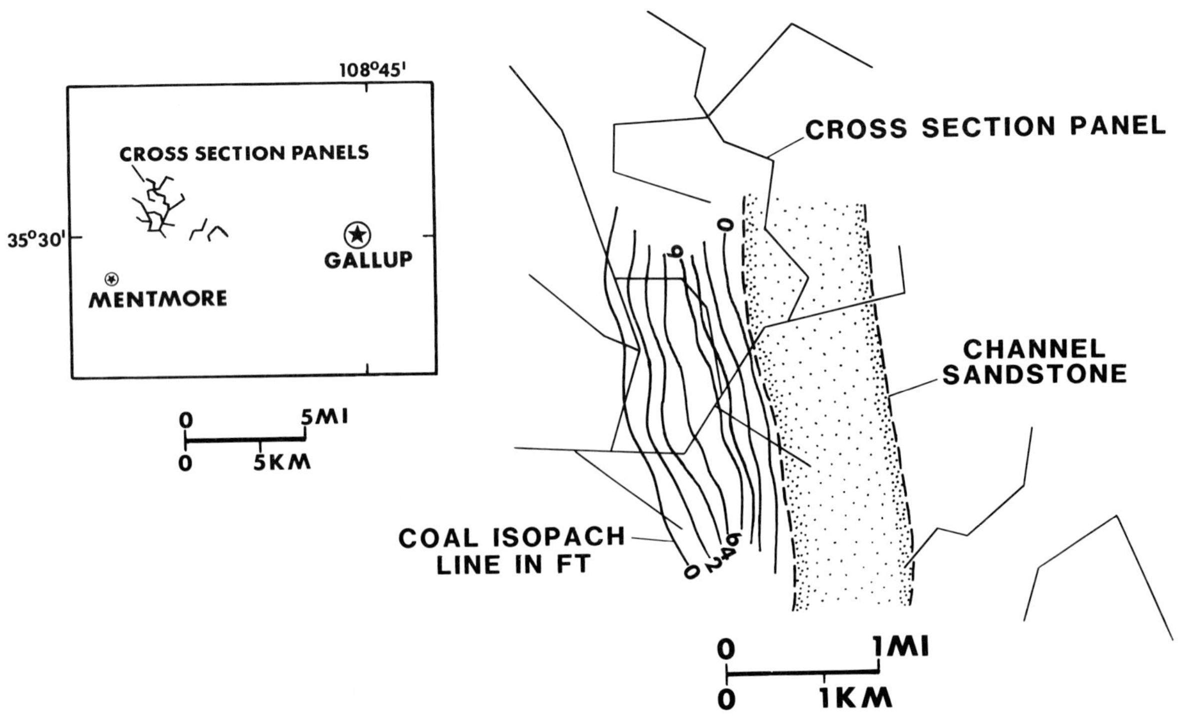

Figure 15. Isopach map of the Enterprise coal showing its areal distribution in relation to contemporaneous channel-levee facies.

levee facies at the east. The coal body is elongate along a north-south direction parallel to the depositional dip of the adjoining channel deposits. In coal exploration and development, the target areas would be along a north-south direction parallel to the elongation of the coal body. In contrast, in hydrocarbon exploration the target areas would be to the east, in the direction of coal pinchout, where the channel-levee facies is developed. Thus, the delineation of the spatial distribution and geometry of the coal beds or source facies assists in predicting and recognizing reservoir facies. This method of extrapolation of reservoir facies is especially applicable in coal-bearing or gas-prone fields.

In the Raton Basin, the megacyclic occurrence of coal-bearing and mud-rich fluvial facies (e.g., lower and upper coal zones) and sand-rich fluvial facies (e.g., quartzose and conglomeratic channel sandstones) in the upper alluvial plain presents an ideal vertical succession of source and reservoir facies under subsurface conditions in gas-prone basin fills. The sand-rich fluvial facies serve as major conduits for fluid migration, and the bounding coal-bearing and mud-rich fluvial facies act as sealing and gas-prone source facies. The internal architecture of the sand-rich fluvial facies varies from heterogeneous, clustered, lenticular sandstone reservoirs segregated by thin floodplain facies at the basin proper to homogeneous, continuous, sheetlike sandstone reservoirs that lack internal segregation by floodplain facies at the basin margin. The sand-rich fluvial facies form a continuum of sandstone reservoirs that thicken toward the basin margin. The dominance of one type of internal architecture to the other in the sand-rich fluvial facies in the basin fill is controlled by uplift of the source area enhanced along thrust faults. The presence of basin-fringe thrust faults provides structural traps that prevent further upward escape of gas. The coal-bearing and mud-rich fluvial facies contain isolated, lenticular (shoestring) reservoirs. These isolated sandstone reservoirs form linear meanderbelts along depositional dip, which are bounded by abandoned channel-plug and cutbank-slump deposits. In addition, thin, locally lobate crevasse-splay sandstone interlaced by branching crevasse-channel sandstone constitutes a volumetrically small but equally important reservoir. The source and sealing facies are more common in the coal-bearing and mud-rich fluvial facies than in the sand-rich fluvial facies. Thus, ideally, isolated sandstone and crevasse-splay reservoirs, although more difficult to delineate than those of the sand-rich fluvial facies, may be potentially more productive reservoirs because of close proximity to gas-prone source facies (e.g., coal beds).

Dolly and Meissner (1977) reported the occurrence of gas in the Raton Basin, which was derived principally from thermal generation of coal beds. According to Dolly and Meissner, coal deposits, which consist of a low hydrogen-to-carbon ratio, initially generate large quantities of dry methane gas at high volatile B bituminous stage that contains 41 percent volatile matter and 13,500 Btu (moist free); this stage represents the "maturity' threshold (Fig. 16). Generation of gas is beleived to be directly related to devolatilization of coals from a high volatile C bituminous coal to higher rank coals (Francis, 1961). Gas is generated beyond the threshold of maturity; however, it is much less except for the methane generated from anthracite coal.

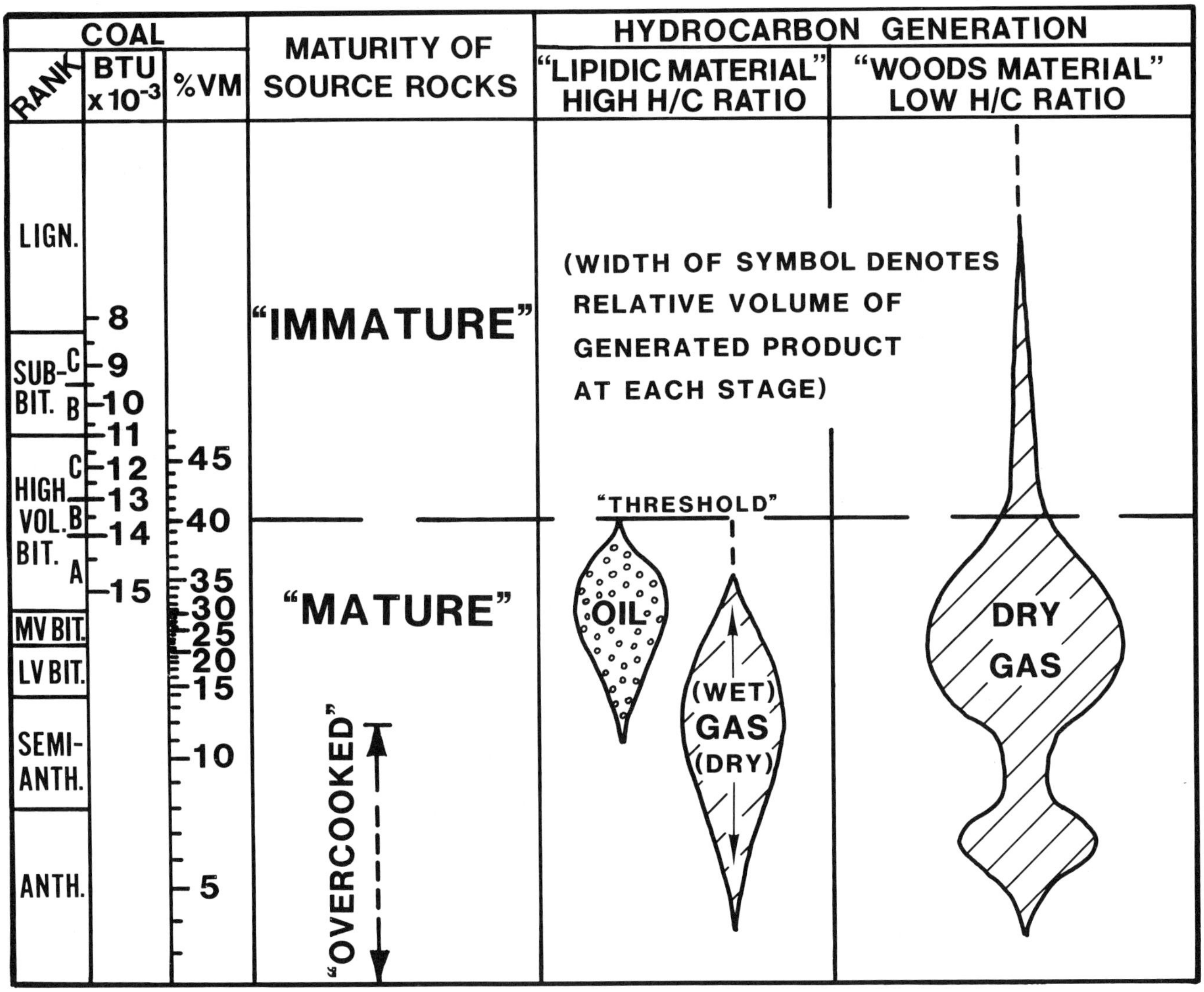

Figure 16. Threshold of significant generation of hydrocarbon from two major types of organic material relative to coal rank. Modified from Dolly and Meissner (1977).

Coal beds of the Raton Formation, as much as 12 ft thick, were interpreted by Dolly and Meissner (1977) as "mature" to "near mature" with respect to large scale methane generation around the basin margins and at an "advanced mature" stage in deeper and central parts of the basin. Figure 17 shows a "maturity" map of the Raton coal beds in the basin based on published analyses of coal rank and volatile matter by Dolly and Meissner (1977) and Amuedo and Bryson (1977). Coal ranks in the Raton Formation at the outcrops and shallow mines are high volatile A and B bituminous in the north-central part of the basin. A contour of equal volatile content (40 percent) of the Raton coals indicates that approximately a 300 sq-mi area is underlain by "mature" coals of high volatile B bituminous or higher rank at the central and deeper parts of the basin. Raton coals along the Purgatoire River, which are in close proximity to the structural basin axis, are at the threshold of high volatile A bituminous coals. Thus, this part of the basin holds some potential for shallow gas accumulation in stratigraphic traps (e.g., channel and crevasse-splay sandstones).

In the Powder River Basin, the most important reservoir facies in the intermontane alluvial plain consist of isolated, linear meanderbelts of trunk and tributary channel sandstone and crevasse-splay sandstone. Internally, the meanderbelt sandstone is a heterogeneous reservoir locally disconnected by clay and silt drapes of lateral accretion units. The tributary meanderbelt sandstone is more restricted in dimension than the trunk meanderbelt sandstone. These sandstone reservoirs have branching outlines; the trunk sandstone forms a linear belt oriented with depositional dip, and the tributary sandstone develops transverse to the depositional dip. Tributary sandstone, which consists of bedload sediments

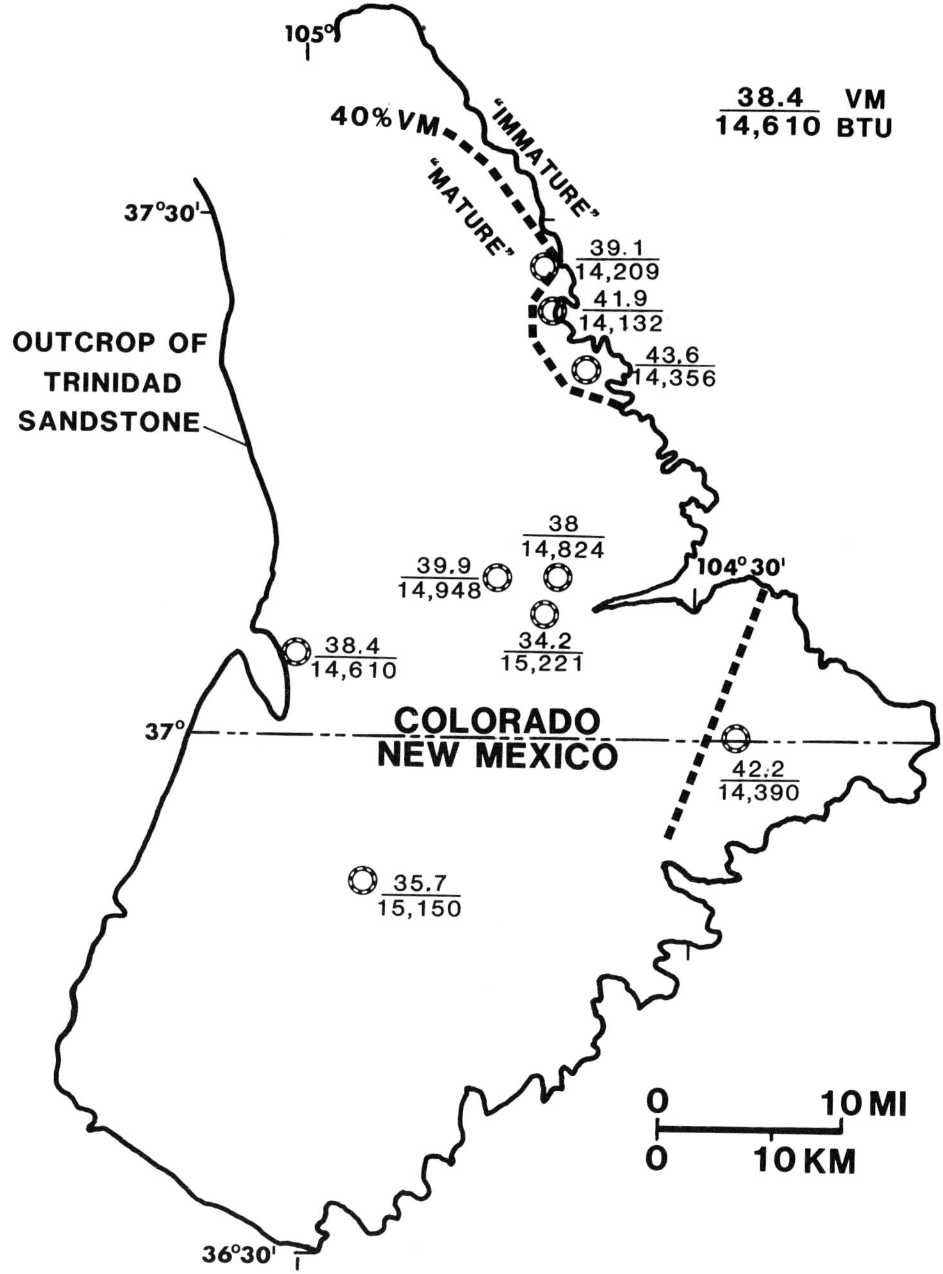

Figure 17. Map showing maturity of Raton coals in the Raton Basin. Modified from Dolly and Meissner (1977).

and merges with alluvial fan deposits toward the basin fringe, may form more permeable and transmissive reservoirs than the trunk sandstone. The trunk sandstone was deposited either by meandering or anastomosing streams, which are mixed-load and suspended-load fluvial systems, respectively. The crevasse-splay sandstones form as erratic lobes or tongues that extend and thin from the belts of trunk and tributary sandstones. These isolated crevasse wedges can form small economic reservoirs proximal to major channel-fill reservoirs.

The sealing facies of the intermontane fluvial facies includes levee, floodplain, lake, and backswamp muds. These fine-grained deposits contain considerable amounts of organic debris derived from woody and herbaceous plants and layered coal beds that are gas-rich source facies. The coal source facies is thicker and more laterally extensive than the coal of the coastal plain-piedmont alluvial plain; therefore, coal of the intermontane alluvial plain makes an excellent mappable unit that may assist in delineating the spatial distribution of reservoir facies. The Felix coal, for instance, covers as many as 1000 sq mi and is much more extensive than coals in the coastal plain-piedmont alluvial plain of the Gallup sag and the Raton Basin (as much as 25 sq mi).

The coal-bearing Fort Union and Wasatch Formations in the Powder River Basin are potentially important sources of coal-derived biogenic gas. Law (1984) reported that gas seeps, flowing gas wells, and gas shows in shallow drill holes suggest that these rock units contain economically recoverable methane gas resources. Analyses of the chemical and isotopic properties of coal-derived gas and gas recovered from sandstone reservoirs suggest that they are identical and biogenic ($\delta^{13}C$ values range from -53.59 to -60.85 percent and C_1-C_{1-5} values range from 0.97 to 0.99) (Law, 1984). Table 1

Table 1. Shallow biogenic gas production in the Wasatch and Fort Union Formations in the Powder River Basin (from Law, 1984).

	Moose Draw Field	Unnamed Field	Hay Creek Field*
Formation	Wasatch	Tongue River Member of the Fort Union Formation	Tongue River Member of the Fort Union Formation
Number of Wells	4	2	2
Depth	250-300 Ft	444-516 Ft	223-415 Ft
Cumulative Production	58 mmcf (9/80-2/84)	127.6 mmcf (2/81-9/83)	1-2.6 mmcf/d** (7/34)

* Abandoned
** Initial production

shows shallow gas production from 3 fields in the Powder River Basin where production is from the Tongue River Member of the Fort Union Formation and the Wasatch Formation at depths of 223 to 516 ft. Cumulative production is as much as 127 million cu ft/day. Structural trap (e.g., compaction folds) and stratigraphic trap (e.g., channel sandstone) are in stratigraphic proximity to gas source rocks (coal) providing maximum conditions for early entrapment of biogenic gas (Law, 1984).

Thus, evaluation of fluvial facies in alluvial plains of the coastal plain-piedmont setting and in intermontane alluvial plains of the Rocky Mountain region reveals a variety of fluvial styles of deposition. A common denominator in all of these fluvial facies is the presence of coal. Understanding the interrelationships of the physical characteristics of coal deposits and genetic facies of associated sediments provides a key to the exploration and development of economic coal beds. The parameters used to explore and develop economic coal bodies can also be adopted to predict, recognize, and delineate stratigraphic traps for hydrocarbons.

SUMMARY

Alluvial plain environments landward of Late Cretaceous maximum transgressive coastlines and Late Cretaceous-early Tertiary regressive coastlines, and those in Tertiary intermontane settings, are representative of contrasting coal-forming environments in the Rocky Mountain region. The lower alluvial plain and its coal-forming environments in the Gallup sag are directly influenced by regional transgression of coastal lagoons accompanied by maximum landward advance of coastal barriers. The lower alluvial plain backswamps were sustained by a rise of the ground-water table due to lagoonal transgression over randomly abandoned fluvial facies. In the Raton Basin, the upper alluvial plain and its coal-forming settings are affected by extrabasinal and intrabasinal tectonism, which controlled the megarhythmic occurrence of coal-bearing intervals and the intervening sand-rich sequences of alluvial fan, braided stream, and high-bedload meandering stream deposits. In the Powder River Basin, the intermontane alluvial plain and its coal-forming environments are also controlled by tectonism. Alluvial fans shed coarse-grained sediments at the western basin margin and also served as areas of recharge for the ground water that sustained the backswamps. In addition, the Powder River and Raton Basins both formed as a dynamic compacting basins, and their compaction influenced the upward motion of deep-basin ground water, which in turn supported backswamps of the alluvial plain.

REFERENCES

ALLEN, J. R. L., 1963, The classification of cross stratified units, with notes on their origin: Sedimentology, v. 2, p. 93-114.

ALLEN, J. R. L., 1965, A review of the origin and characteristics of recent alluvial sediments: Sedimentology, v. 5, p. 89-191.

AMUEDO, C. L., AND BRYSON, R. S., 1977, Trinidad-Raton Basins: A model coal resource evaluation program, in Murray, D. K., ed., Geology of Rocky Mountain Mountain Coal, Proceedings of the 1976 Symposium: Colo. Geol. Survey Res. Ser. 1, p. 45-60.

BLACKSTONE, D. L., JR., 1981, Compression as an agent in deformation of the east-central flanks of the Bighorn Mountains, Sheridan and Johnson Counties, Wyoming, in Boyd, D. W., and Lillegraven, J. A., eds. Rocky Mountain Foreland Basement Tectonics: Contrib. Geology, Univ. Wyo., Laramie, v. 19, p. 105-122.

CAVAROC, V. V., AND FLORES, R. M., 1984, Lithologic relationships of the Upper Cretaceous Gibson-Cleary stratigraphic interval, in Rahmani, R. A., and Flores, R. M., eds., Sedimentology of Coal and Coal-Bearing Sequences: Intern. Assoc. Sediment., Spec. Pub. 7, p. 197-216.

DOLLY, E. D., AND MEISSNER, F. F., 1977, Geology and gas exploration potential, upper Cretaceous and lower Tertiary strata, northern Raton Basin, Colorado, in Veal, H. K., ed., Exploration Frontiers of the Central and Southern Rockies: Rocky Mtn. Assoc. Geol. Symp. 1977, Denver, Colo., p. 247-270.

FLORES, R. M., 1981, Coal deposition in fluvial paleoenvironments of the Paleocene Tongue River Member of the Fort Union Formation, Powder River area, Powder River Basin, Wyoming and Montana, in Ethridge, F. G., and Flores, R. M., eds., Recent and Ancient Nonmarine Depositional Environments--Models for Exploration: Soc. Econ. Paleon. Mineral. Spec. Paper No. 31, p. 169-190.

FLORES, R. M., 1984, Comparative analysis of coal accumulation in Cretaceous alluvial deposits, southern United States Rocky Mountain basins, in Stott, D. F., and Glass, D. J., eds., The Mesozoic of Middle North America: Can. Soc. Petrol. Geol. Mem. 9, p. 373-385.

FLORES, R. M., AND TUR, S. M., 1982, Characteristics of deltaic deposits in the Pierre Shale, Trinidad Sandstone, and Vermejo Formation, Raton Basin, Colorado: The Mtn. Geologist, v. 19, p. 25-40.

FLORES, R. M., AND WARWICK, P. D., 1984, Dynamics of coal deposition in intermontane alluvial paleoenvironments, Eocene Wasatch Formation, Powder River Basin, Wyoming, in Houghton, R. L., and Clausen, C. N., eds., Proceedings 1984 Symposium on Geology of Rocky Mountain Coal: North Dakota Geol. Soc., Pub. 84-1, p. 184-199.

FRANCIS, W., 1961, Coal, its formation and composition: Edward Arnold Ltd., London, 806 p.

KREITLER, C. W., 1979, Ground-water hydrology of depositional systems, in Galloway, W. E., Kreitler, C. W., and McGowen, J. H., eds., Depositional and Ground-water Flow Systems in the Exploration of Uranium: Bur. Econ. Geol. Res. Colloquium, Univ. Texas, Austin, p. 118-176.

LAW, B. E., 1984, Biogenic bas accumulations in large-scale compaction structures, Powdser River Basin, Wyoming and Montana (abs.): Am. Assoc. Petrol. Geol. Bull., v. 68, p. 940.

PILLMORE, C. L., AND FLORES, R. M., 1984, Coal deposits, stratigraphy, and the Cretaceous-Tertiary boundary, southern Raton Basin, New Mexico and Colorado (Field Trip 3), in Lintz, J., Jr., ed., Western Geological Excursion: Dept. Geol. Sci., Mackay Sch. Mines, Reno, Nevada, v. 3, p. 1-51.

WOODWARD, L. S., AND SNYDER, D. O., 1976, Structural framework of the southern Raton Basin, New Mexico: New Mex. Geol. Soc. Guidebook, 27th Field Conf., Vermejo Park, p. 125-127.

CHAPTER 9

RESERVOIR CHARACTERISTICS OF ANCIENT FLUVIAL DEPOSITS WITH EMPHASIS ON ROCKY MOUNTAIN AND MID CONTINENT REGIONS

Frank G. Ethridge
Department of Earth Resources
Colorado State University

Introduction

Fluvial depositional systems often constitute significant host for hydrocarbons. In the Frio Depositional Systems in Texas, the Gueydan fluvial system has the highest whole rock yield factor of all Frio Systems (Galloway, et al., 1982) and in Prudhoe Bay field, Alaska, the largest oil field in North America, braided stream deposits host the principal producing reservoir. Interest in hydrocarbon-bearing, fluvial depositional systems remains high, especially with regard to the Cretaceous in the Powder River Basin, Wyoming,the Tuscaloosa in Mississippi and the Pennsylvanian in the D-J Basin, Colorado. Because of the diverse nature of fluvial deposits, the characteristics of hydrocarbon-bearing fluvial reservoirs are variable. At one extreme are braided river deposits, which have abundant potential reservoir rocks but lack seals, and at the other extreme are fine-grained meander belt deposits with small- to moderate-size reservoirs isolated in mudrock (Galloway and Hobday, 1983).

In this chapter we will review the characteristics of point bar, braided stream and fan delta reservoirs and data pertaining to the recognition of paleotectonics, unconformities and hydrocarbon-bearing valley fill sequences. Special emphasis will be placed on an evaluation of fluvial hydrocarbon-bearing reservoirs in the Rocky Mountains and the Mid-Continent.

Fine-Grained Meanderbelt Systems

Point bar deposits, which result from deposition on the concave bank of a meandering stream during falling river stage following flooding, constitute the principal reservoirs in fine-grained meander belt systems (Fig. 1). Point bar deposits are recognized in vertical sequences by an erosional base, an overall decrease upward in the scale of sedimentary structures from trough cross-beds and horizontal laminations to climbing ripple stratification, and by a decrease upward in grain size, which may be reflected by the typical bell-shaped log response of SP or GR curves (Fig.9, Chapter 2). Point bar deposits in

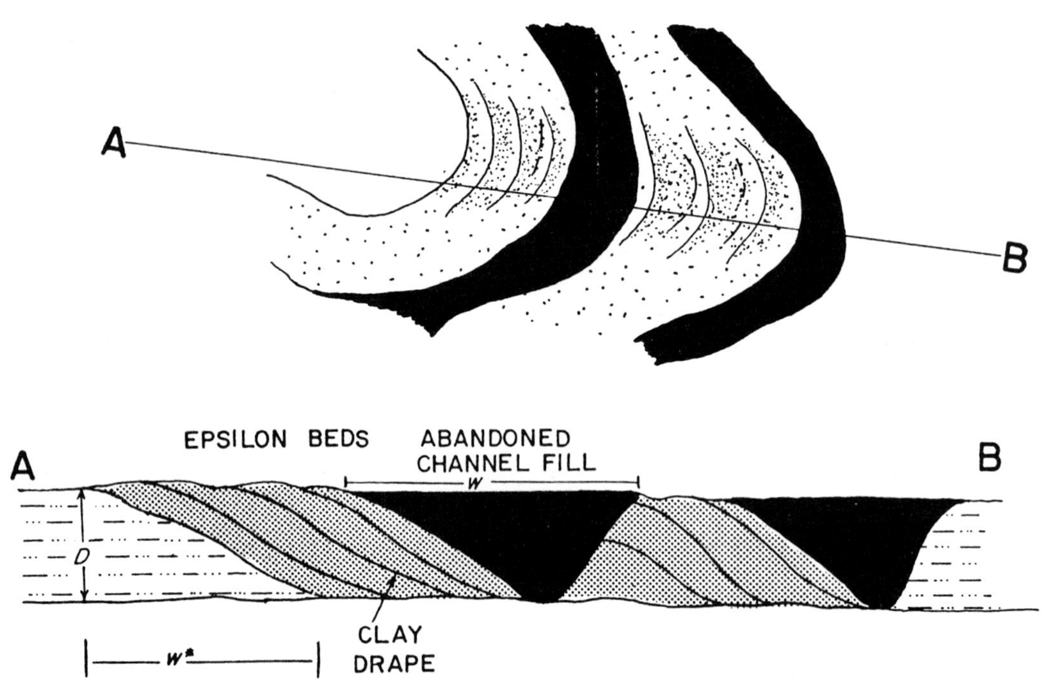

Figure 1. (A) Plan view map of hypothetical meander bend showing location of cross section in (B). (B) cross section of hypothetical lateral accretion depoits illustrating large-scale sigmoidal cross-stratified units (epsilon beds), depth and width measurements used in calculation of paleo channel morphologic and flow parameters. Note presence of inhomogenetics represented by abandoned channel fill deposits and clay drapes.

general have a discontinuous sheet or beaded belt and ribbon geometry in plan and cross section view respectively (Fig. 5., Chapter 2). Individually point bars might display an arcuate or ovoid patterns (Cornish, 1984). Deposits generally trend at high angles to depositional strike. Internally point bar deposits are extremely heterogeneous and reveal at least three orders of inhomogeneity. The first is within individual epsilon beds and is related to random variations in grain size and sediment type. The second, clay drapes (Fig. 1), are found between successive epsilon beds and result from deposition of fine-grained sediment during falling river stage. The third, abandoned channel-fill deposits (Fig. 1), are related to channel avulsion. Traps are commonly styratigraphic and relate to facies changes from point bar sandstones to abandoned channel fill or floodplain deposits or to the edge of the entrenched valley.

Ethridge and Schumm (1978) use of thickness and width of epsilon cross strata (Fig. 1) and the grain-size of channel and bank deposits to determine paleomorphologic parameters (size and shape of point bar deposits). A variation on the technique was recently applied to data on point bar thicknesses from a single drill hole to estimate the width of a sandstone meander belt complex in the Mesaverde Complex, Pieance Creek basin, Colorado (Lorenz, Heinze, Clark and Searls, 1985). The technique allows a "ball-park" estimate of field size after only one well has been drilled. More commonly the size and shape of point bar reservoirs are determined after a number of wells have been drilled and parameters such as channel width (W), meander wave length (ML), radius of curvature (RC) and meander belt width (MBW), can be measured directly

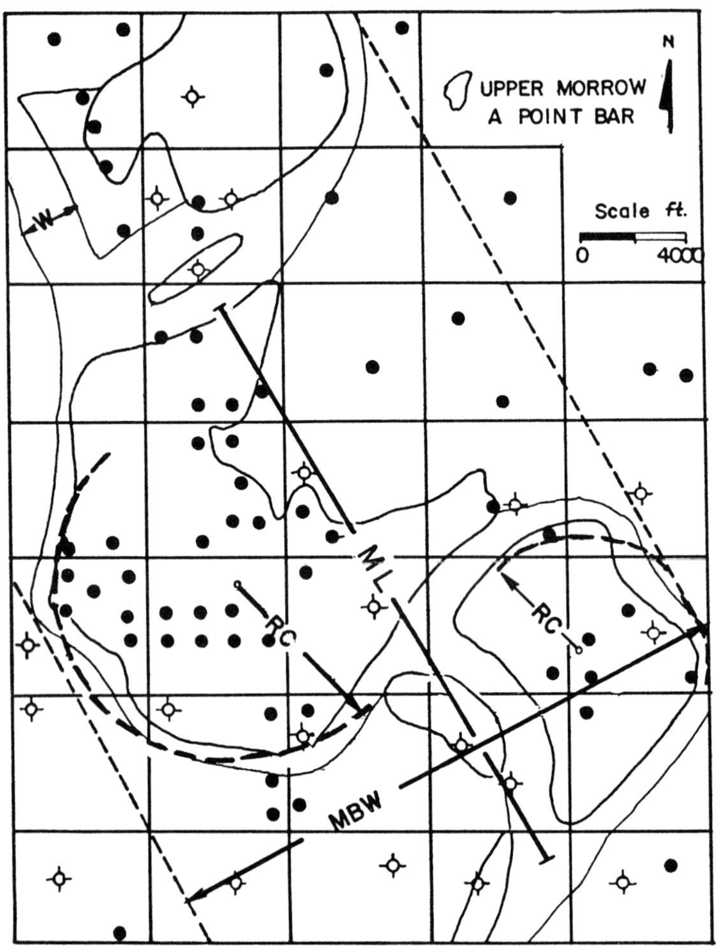

Figure 2. Typical meander belt parameters (stream width (W), meander wavelength (ML), radius of curvature (RC), and meanderbelt width (MBW) that can be determined given sufficient wireline log and/or continuous core data. Upper Morrow A (Pennsylvanian) Sandstone, Beaver County, Oklahoma (modified from Cornish, 1984).

from plan view reconstructions (Fig. 2; Berg, 1968, Ebanks and Weber, 1982, and Cornish, 1984). Parameters determined from these studies along with parameters determined by this author from other fields permit an assessment of the size range of individual point bar reservoirs in the mid-continent and Rocky Mountain regions of the U.S. The reservoirs examined are of Paleozoic and Mesozoic age and include the Pennsylvanian Morrow Sandstone of the Oklahoma Panhandle and southeast Colorado, the Cherokee Sandstone of Kansas, the Cretaceous J Sandstone of eastern

Colorado and the Fall River and Muddy Sandstones of eastern Wyoming, and Cretaceous sandstones in western Colorado and Mississippi (Table 1). Sizes range from the small Cherokee meander belt system (Fig. 3A) with 8 individual point bar reservoirs across a 1 mile square section (Fig. 3B), which is comparable in size to the modern Laramie River, Wyoming (Fig. 3C), to the moderately large Coyote Creek reservoir of the Fall River trend along the eastern side of the Powder River Basin, Wyoming (Fig. 4). The reservoir at Coyote Creek field covers portions of five, 1-mile square sections and is comparable in size to the Holocene point bar deposits along the middle Mississippi River.

Valley-Fill Sequences

The importance of paleotectonics and its role in controlling sedimentation patterns and hydrocarbon accumulations in the Rocky Mountain region has received considerable attention in recent years (Slack, 1981, Weimer, et al., 1982, Weimer, 1983). Some of these hydrocarbon-bearing deposits, which are now interpreted as point bars in valley-fill sequences, had previously been interpreted as delta distributary channels. Notable exceptions which include the Upper Cretaceous, Cut Bank Sandstone at Cut Bank field, Montana (Blixt, 1941) and the Lower Cretaceous J Sandstone in western Nebraska (Harms, 1966, Fig. 5) were recognized early as valley-fill sequences. Other sequences that have been recently inturpeted as fluvial, valley-fill sequences include the Lower Cretaceous Newcastle Formation in the Fiddler Creek and Clareton hydrocarbon trends in the southeastern Powder River Basin, Wyoming (Weimer, et al., 1982; Fig. 6) and the J Sandstone in the western D-J Basin, Colorado (Weimer, 1983); Fig. 7). The Recluse trend in the northeastern portion of the Powder River Basin, Wyoming that

Table 1. Paleomorphologic and reservoir characteristics of ancient point bar deposits from Wyoming, Colorado, Nebraska, Oklahoma, Kansas and Mississippi (Data from or calculated from Harms, 1966; Berg, 1968; Berg and Cook, 1968; Curry and Curry, 1972; Ebanks and Weber, 1982; Cornish, 1984; Orchard and Kidwell, 1984; Lorenz, Heinze, Clark and Sears, 1985.

FALL RIVER WY	L. TUSCALOOSA MISS	L. MUDDY (NEWCASTLE) WY	U.MORROW OK	U.MORROW CO	WILLIAMS FORK (MESAVERDE) CO	J-SS NEB	CHEROKEE KA	CHARACTER
27500	22780	22440	21900	15840	1600	1500	900	MBW (FT)
33000	22110	35250	24600	16900	----	----	1300	ML (FT)
6250	6125	6800	5800	4224	----	----	250	RC (FT)
1500	2650	2805	2032	1000(?)	225	----	100	W (FT)
70	55	45	48	40+	11.5	30	17.5	D (FT)
20	35	--	3.5	10	--	5	2	R/P(MMB)
19	27	--	15	19	--	20	25	POR (%)
230	5-250	--	200	1-4000	--	350	460	PERM(MD)
OLD COURSE MID. MISS.	U. MISS						LARAMIE	MODERN ANALOG

NOTE: MDW = meander belt width; WL = meander wave length; RC = radius of curvature; W = width; D = depth; R/P = reserves/production (comparisons not directly comparable); PDR = porosity; PERM = permeability (FT) = feet; (MMB) = million barrels; (MD) = millidarcies.

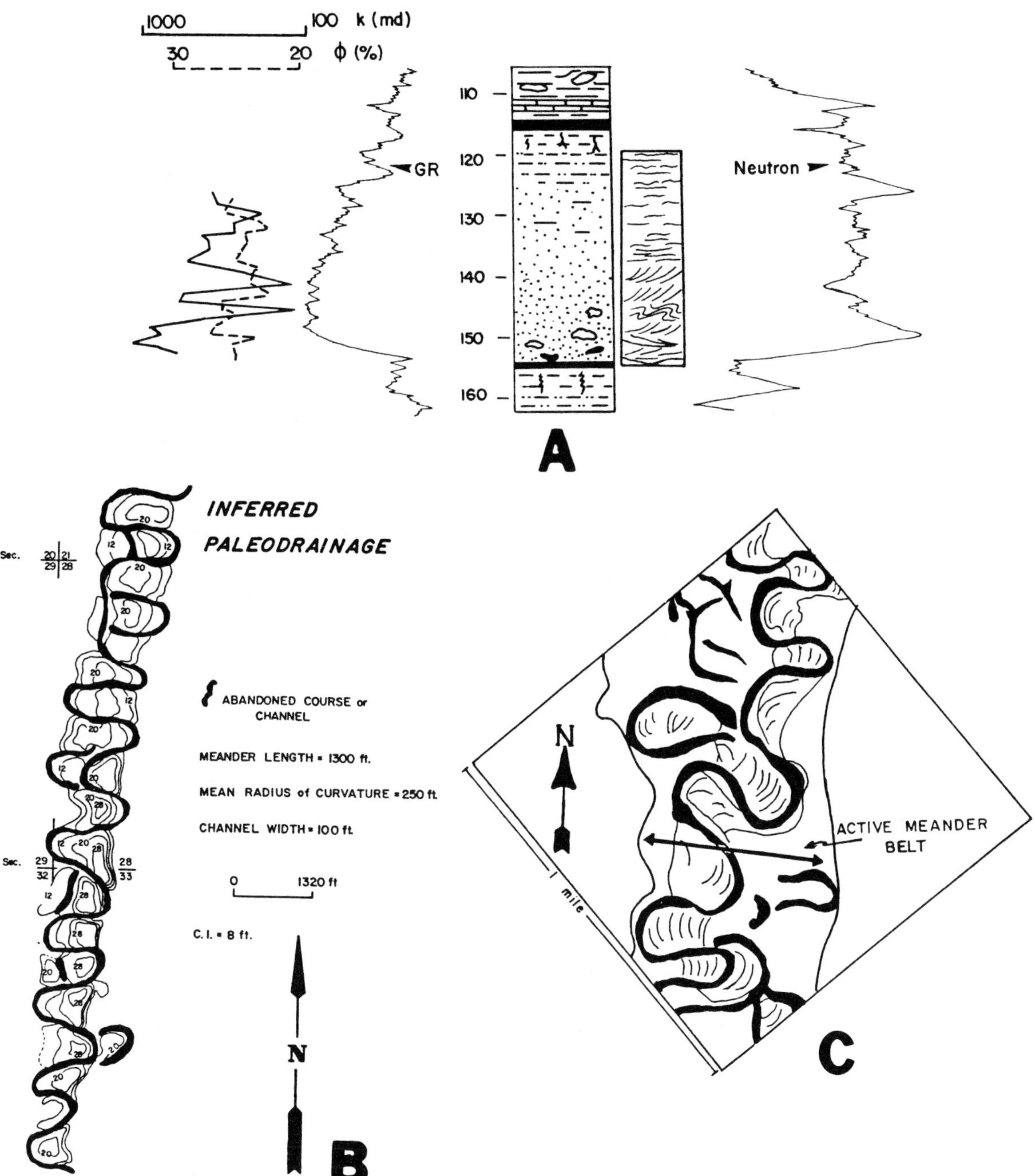

Figure 3. (A) Vertical sequences of lithologies, sedimentary structures, permeabilities, porosities and gamma ray and neutron log profiles, Eastburn Sandstone, Eastburn field, western Missouri. (b) Gross thickness map and inferred abandoned channel fill deposits Eastburn Sandstone, Eastburn field. (C) Diagrammatic outline of the modern Laramire River, Wyoming and associated point bar deposits. (A & B modified from Ebanks and Weber, 1982; C modified from Swanson, 1976).

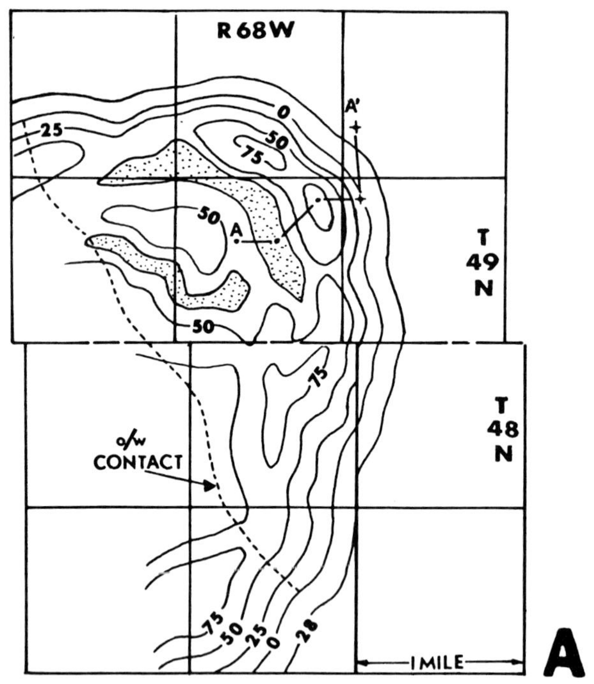

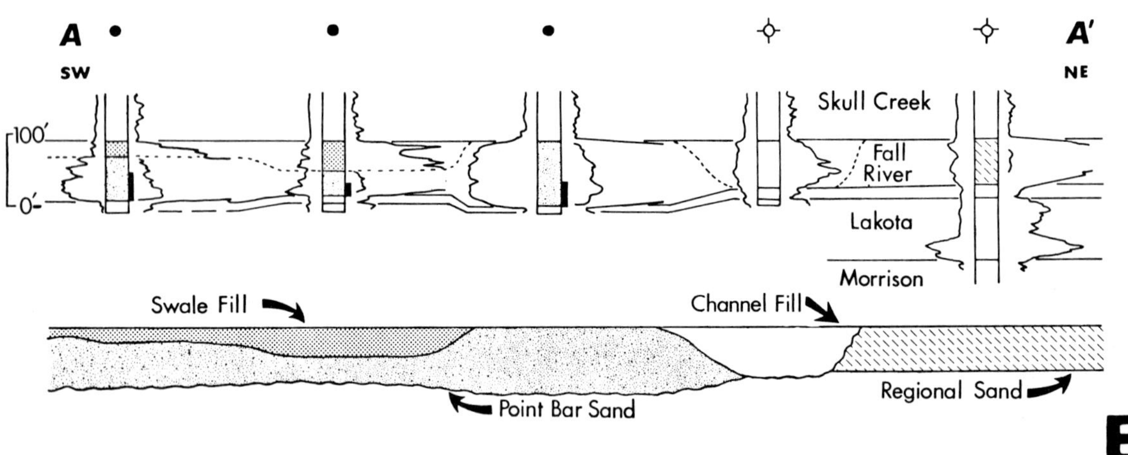

Figure 4. (A) Isopoch map of porous Fall River Sandstone, Coyote Creek field, eastern Powder River Basin, Wyoming. (B) electric log section from Miller Creek field and intepretation of meander belt facies. Location of section shown on A (modified from Berg, 1968).

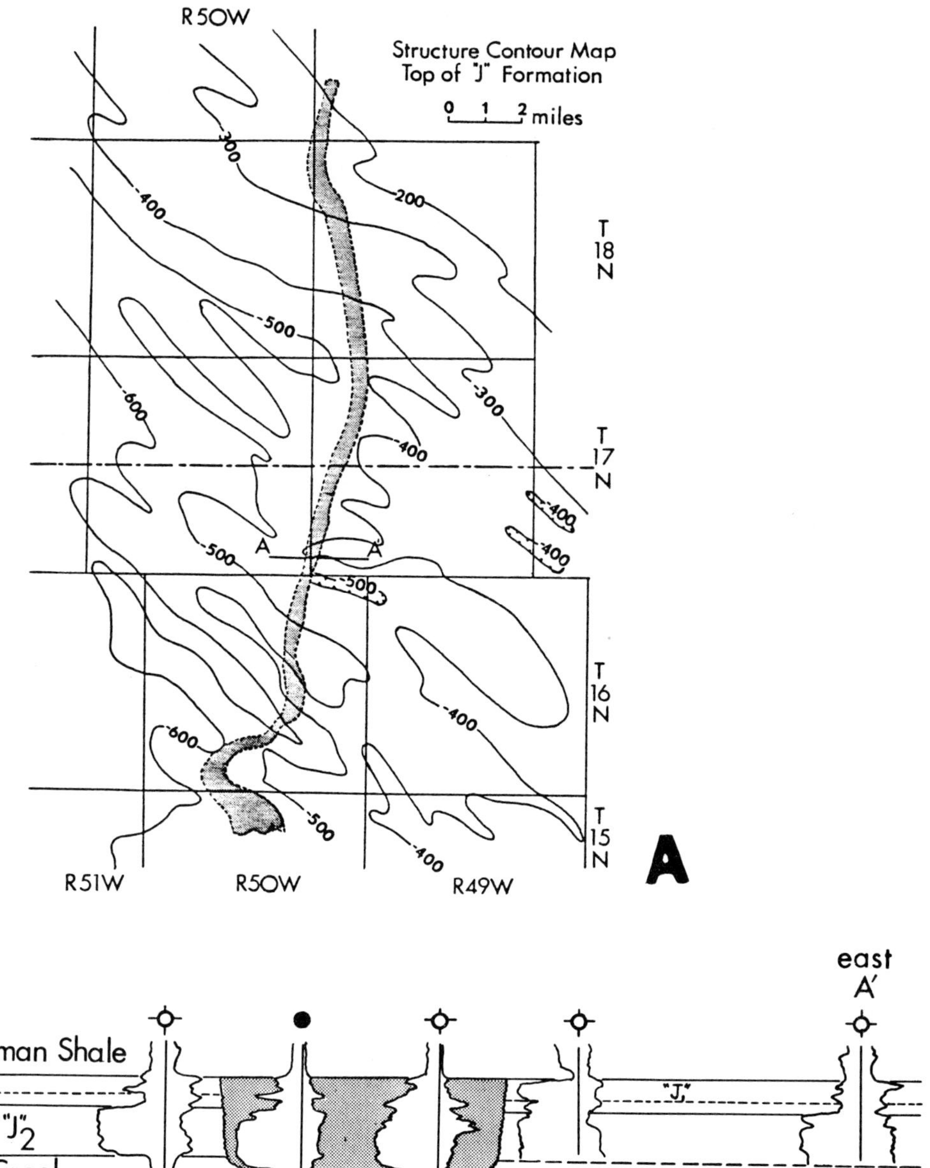

Figure 5. (A) Structural contour map on top of Cretaceous J Sandstone, western Nebraska and location of inferred valley fill sediments. (B) Cross-sections showing inferred valley-fill sequences, J. Sandstone. Note: horizontal scales within and outside inferred valley-fill sequences not the same (modified from Harms, 1966).

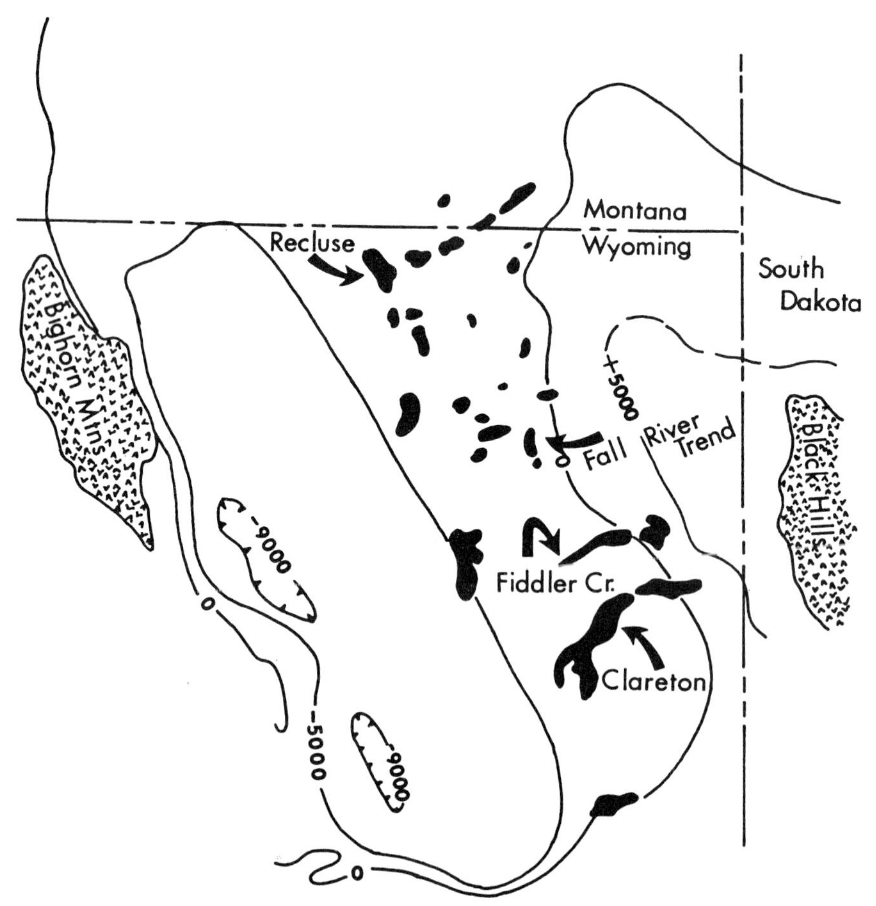

Figure 6. Structure contour map of Powder River Basin, Wyoming showing location of Fiddler Creek, Clareton and Recluse hydrocarbon-bearing trends. Structural contours on top of Lower Cretaceous; contours interval 5000 feet. (modified from Weimer, et al., 1982).

produces from the Lower Cretaceous Fall River Sandstone might have a similar origin (Fig. 6; Stone, 1972 and Woncik, 1972).

Criteria used to recognize valley-fill sequences associated with sea level changes include: 1) inferred shoreline deposits with incised drainage overlain by marine shale; 2) root zones and paleosols near the base of inferred valley-fill sequences; 3) unconformities, recognized by the presence of missing faunal zone, missing facies in a normal regressive sequence, paleokarst, or concentrations of phosphate nodules,

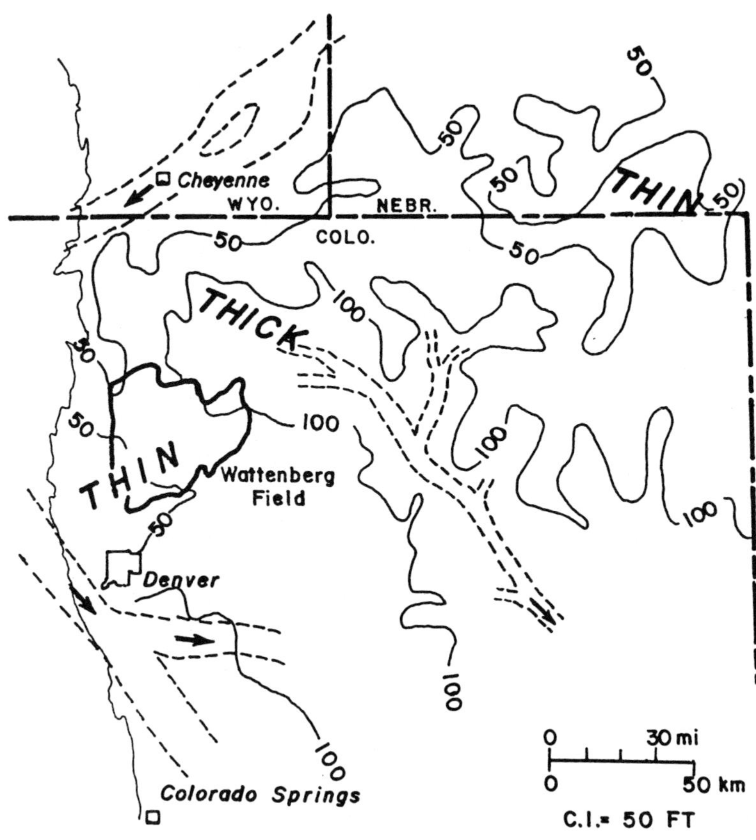

Figure 7. Lower J Sandstone Channel pattern, D-J Basin, eastern Colorado (modified from Weimer, 1983).

glauconite or shell debris; 4) presence of a widespread coal layer overlying marine regressive delta front deposits; and 5) correlation with the record of sea level change from other continents (Weimer, 1983). The importance of correctly identifying these fluvial/deltaic sequences is illustrated by the reinterpration of the J-Sandstone in eastern Colorado (Fig. 7). In this interpretation the J is inferred to be a southeast flowing fluvial system with tributaries. Earlier interpretations had the J Channel system as a northwest-flowing delta distributary system. A valley-fill origin has been considered for the Pennsylvanian, Upper Morrow reservoir at Sorrento field in eastern Colorado and should be considered for Elmwood and other fields along the

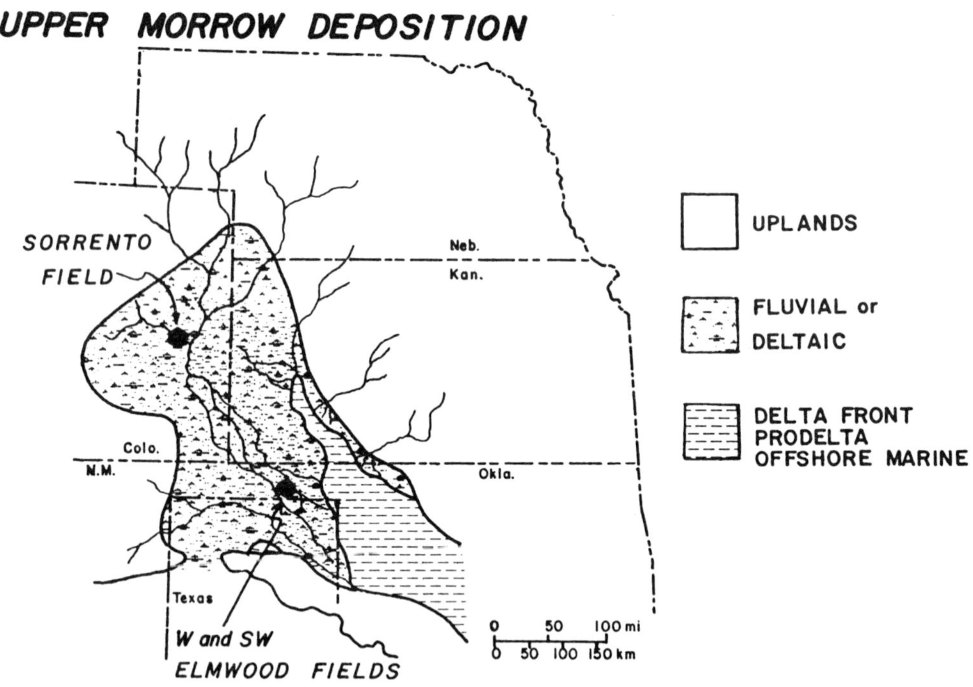

Figure 8. Paleogeographic reconstruction of Upper Morrow time, mid-continent. Note location of Sorrento field in eastern Colorado modified from Swanson, 1979).

same general trend in southeastern Colorado and Oklahoma (Fig. 8; Swanson, 1979 and Orchard and Kidwell, 1984, Sonnenberg, 1985).

Braided Stream Systems

Braided river sandstones and conglomerates are recognized by the presence of planar and/or trough cross stratification, and horizontal stratification and on the basis of cylindrical wireline log patterns that often reflect the lack of an overall grain-size trend (Fig. 9). Braided river deposits are typically more homogeneous than point bar deposits and they display a sheet-like geometry (Fig. 5, Chapter 2; Campbell, 1976) trending normal to depositional strike. Limited stratigraphic trap opportunities exist and most reservoirs result from structural and/or regional unconformity traps, as exemplified by Messla

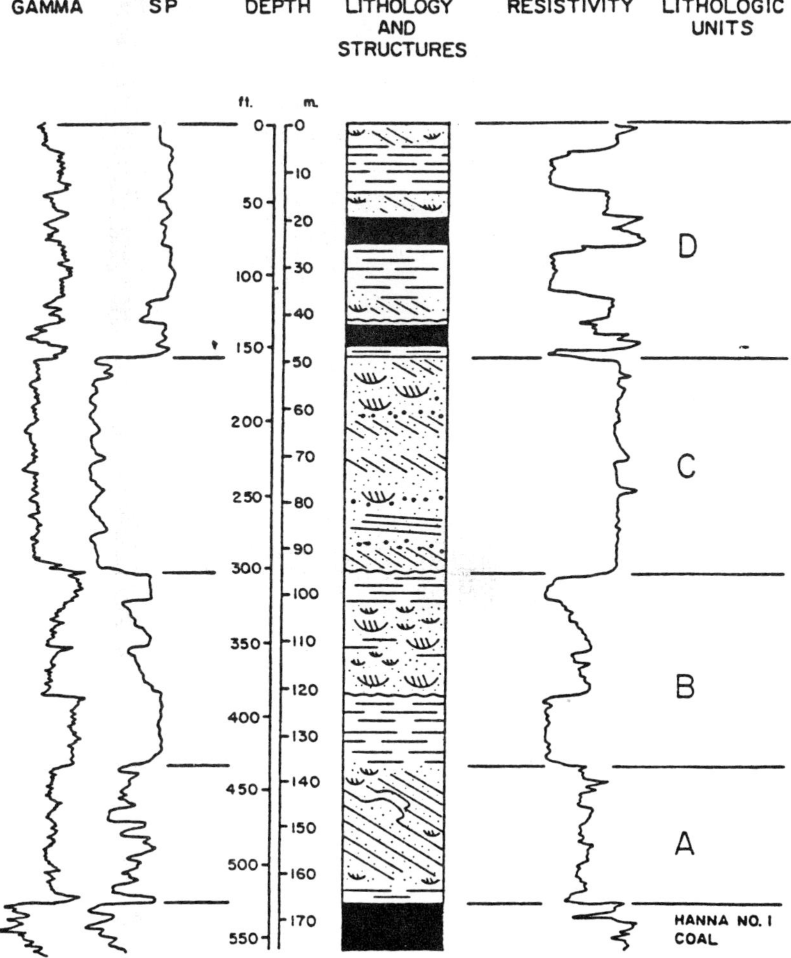

Figure 9. Log patterns and core descriptions of fluvial deposits, Hanna Formation, Hanna Basin, Wyoming. Note cylindrical log shape of inferred braided stream deposits of unit C (150 to 300 feet) (after Craig, et al., 1982).

field, Libya and Prudhoe Bay field, Alaska. These giant fields have ultimate recoverable hydrocarbons in excess of 500 million and 9.6 billion barrels respectively.

Messla field, located on the eastern side of the Sirte Basin (Fig. 10A) is 20 miles long and 4.5 miles wide (Fig. 10B). The Lower Cretaceous Sarir Group, which forms the principal reservoirs at Messla field, consists of two thick sandstones with no preferred grain size trend, separated by a red shale (Fig. 11). Field limits are defined by

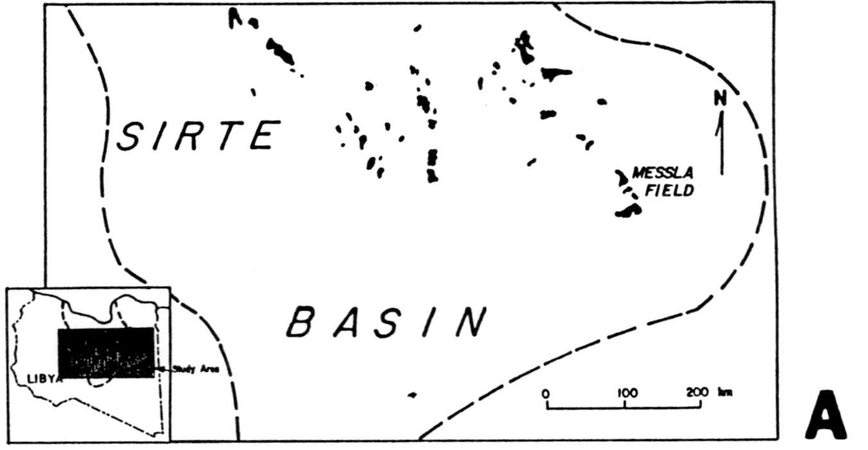

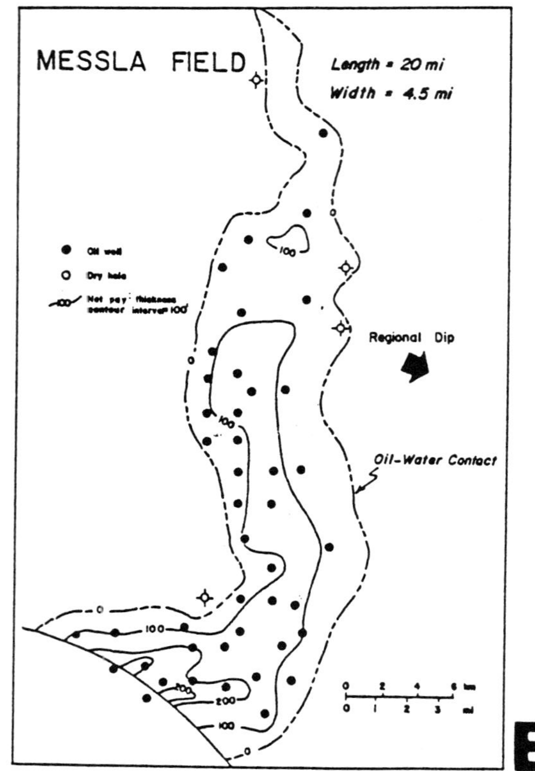

Figure 10. (A) Location map of Messla field, Sirte basin, Libya. (B) Net pay isolith map; Lower Cretaceous, Sarir Santstone Messla field (modified from Clifford, Grund and Musrati, 1980).

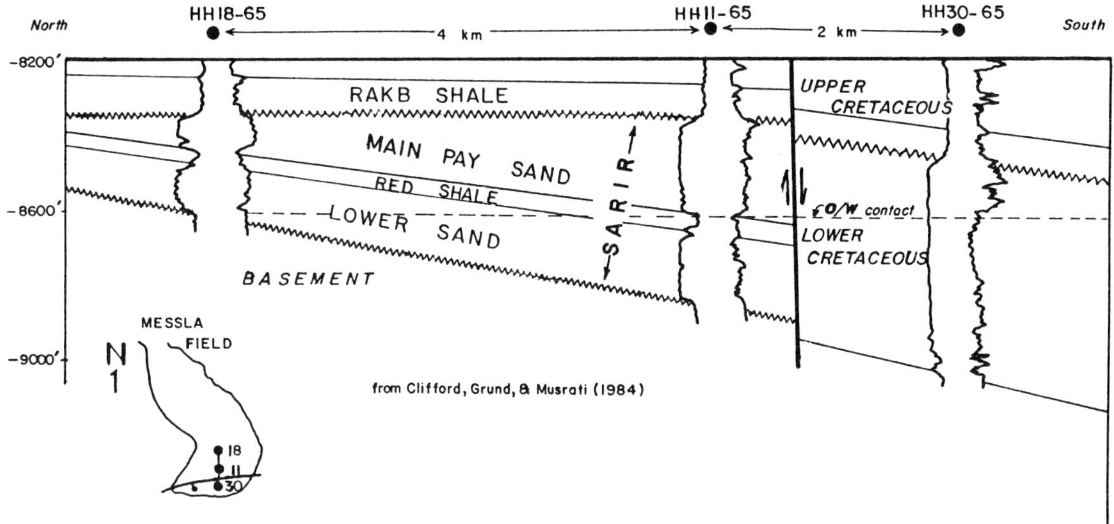

Figure 11. North-south cross section, Messla field, Sirta basin (Note unconformities at top and bottom of Sarir Sandstone) (modified from Clifford, Grund and Musrati, 1980).

the updip erosional limit of the Sarir Sandstone on the west and the oil-water contact on the east. The field's major trapping mechanism is the westerly progression to total erosion and the post Lower Cretaceous unconformity (Fig. 11, Clifford, Grund and Musrati, 1980). Prudhoe Bay field is located on the Barrow Arch, a major subsurface feature on the Artic Slope Province, Alaska (Morgridge and Smith, 1972, Eckelmann, DeWitt and Fisher, 1975; and Jamison, Brockett and McIntosh, 1980). A combined structural-stratigraphic trap related to the Barrow Arch and a major sub-Cretaceous unconformity, which truncates the pre-Cretaceous reservoir down the plunge of the arch provides the trapping mechanism. The principal reservoir rock at Prudhoe Bay is the upper portion of the Lower Triassic, Sadlerochit Formation, which has been interprested as a braided stream complex (Eckelmann, DeWitt and Fisher, 1975; Fig. 12). More recent interpretations of the Sadlerochit Group as a fan-delta complex will be reviewed in the next section.

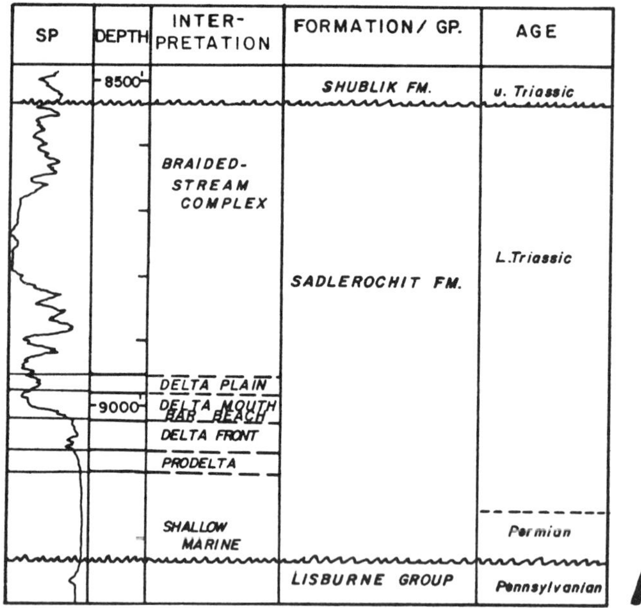

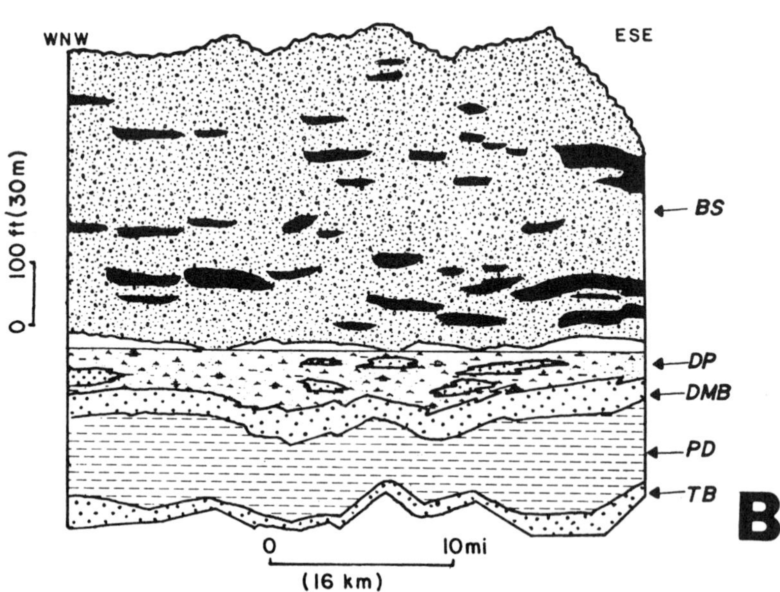

Figure 12. (A) SP curve profile of the principal Sandlerockit Formation units at Prudhoe Bay field, north slope Alaska and inferred environments. (B) WNW-ESE cross section showing distribution lithologies and inferred environments in the Sadlerockit Formation (modified from Eckelmann, DeWitt and Fisher, 1975).

Fan-Delta Systems

Known hydrocarbon-bearing, fan-delta deposits occur principally along divergent plate margins such as the Anadarko basin in the Texas Panhandle and Oklahoma (Sneider, et al., 1977 and Dutton, 1982), the North Sea (Harms, et al., 1982), Brazilian coastal basins (Meister and Aurich, 1972 and Ojeda, 1982), and the Gippsland basin, Australia (Maughan, et al., 1981); and in foreland basins such as the Fort Worth basin, Texas (Tai Wai Ng, 1979) and the Deep basin of Alberta (Leckie and Walker, 1982 and Cant, 1982). Traps in these deposits are mostly of the structural-stratigraphic type. Recently the principal reservoir unit at Prudhoe Bay field, the Ivishak Formation of the Sadlerochit Group, has been reinterpreted as a fan-delta system (Melvin and Knight, 1984 and McGowen and Blockh, 1985). Using the model developed by Wescott and Ethridge (1980, Fig. 13A) Melvin and Knight recognized four distinct lithofacies (L1 to L4, Fig. 13B) in the Ivishak Formation, which they interpreted as shoreline and distal to proximal fan-delta deposits of a shelf-type fan delta.

Ethridge and Wescott (1984) reviewed the characteristics of fan delta reservoirs and generally found that: (1) fan-delta deposits are more important as hydrocarbon reservoirs than previously realized; (2) barring adverse diagenetic effects, delta front conglomerates and sandstones with increased bed segregation and continuity and better sorting need more productive reservoirs, but delta plain, braided channel deposits form the principal reservoirs in some fields; (3) most of the porosity associated with fan-delta deposits is secondary in origin; (4) reservoir potential of most fan-delta deposits is difficult to evaluate with conventional wireline logging devices because of

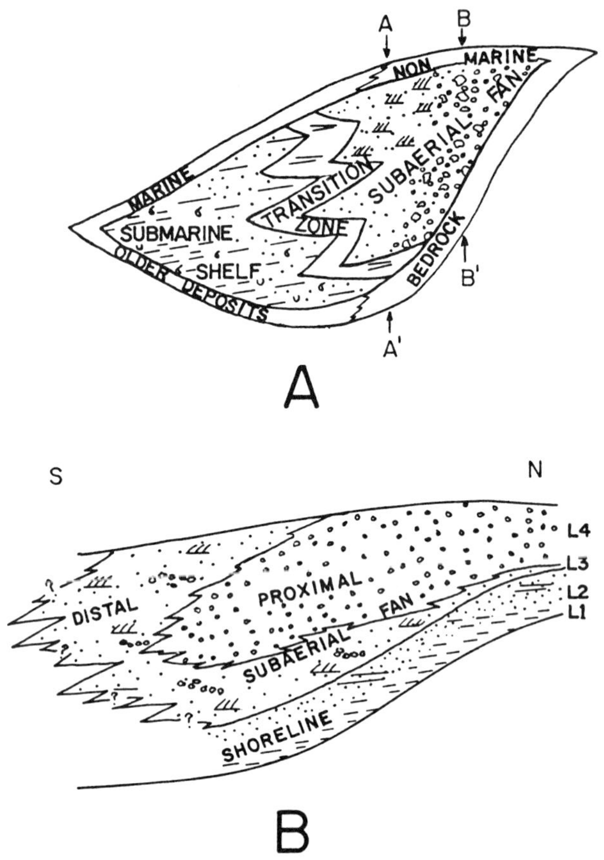

Figure 13. (A) Stratigraphic model for fan deltas built onto continental or island shelf (from Wescott and Ethridge, 1980). Vertical section A-A' is similar to the sedimentary succession in the Ivishak Formation. Vertical section B-B' is not found in sedimentary succession of the Ivishak Formation. (B) Generalized north-south section through the Ivishak Formation at Prudhoe Bay, Alaska. Note lateral persistence of the basal shoreline deposits and the absence of any analogous vertical section (B-B') modified from Melvin and Knight, 1984).

radioactive detrital mineral grains and diagenetic cements commonly present in these immature deposits; and (5) reservoir quality is affected by the bimodal conglomeratic texture of the deposits and by the type and distribution of cements.

REFERENCES

Berg, R.R. 1968, Point-bar origin of Fall River Sandstone reservoirs, northeastern Wyoming: AAPG Bull., v. 52, p. 2116-2122.

Berg, R.R. and Cook, B.C., 1968, Petrography and origin of lower Tuscaloosa sandstones, Mallalieu field, Lincoln County, Mississippi: Trans. Gulf Coast Assoc. Geol. Socs., v. XVIII, p. 242-255.

Blixt, J.E., 1941, Cutbank oil and gas field, Glacier county, Montana, _in_, Levorsen, A.T. (ed.), Stratigraphic type oil fields: Amer. Assoc. Petrol. Geol., p. 327-381.

Broadhead, R.F., 1984, Geology of gas production from 'tight' Abo red beds, east central New mexico: Oil & Gas Journal, June 11, 1984, p. 147-158.

Campbell, C.V., 1976, Reservoir geometry of a fluvial sheet sandstone: Amer. Assoc. Petrol. Geol. Bull., v. 60, p. 1009-1020.

Cant, D.J., 1982, Sedimentology and petroleum geology, Spirit River Formation (Lower Cretaceous), Deep Basin, Alberta (abs): Amer. Assoc. Petrol. Geol. Bull., v. 66, p. 824.

Clifford, H.J., Grund, R. and Musrati, Hassen, 1980, Geology of a stratigraphic giant: Messla oil field, Libya, _in_ Halbouty, M.T. (ed.), Giant oil and gas fields of the decade 1968-1978: AAPG Memoir 30, p. 507-524.

Cornish, F.G., 1984, Fluvial environments and paleo hydrology of the Upper Morrow 'A' (Pennsylvanian) meanderbelt sandstone, Beaver County, Oklahoma: Shale Shaker, V. 34, p. 70-80.

Craig, G.N., II, Burns, L.K., Ethridge, F.G., Laughter, T.F. and Youngberg, A.D., 1982, Overburden characteristics and post-burn study at the Hanna, Wyoming underground coal gasification site: stratigraphy, depositional environment and mineralogy of the Hanna Formation: U.S. DOE/LC/10496-T1 (DE 82009053), 188 p.

Dutton, S.P., 1982, Pennsylvanian fan-delta and carbonate deposition, Mobeetie Field, Texas panhandle: Amer. Assoc. Petrol. Geol. Bull., v. 66, p. 389-407.

Ebanks, W.J., Jr. and Weber, J.F., 1982, Development of a shallow heavy oil deposit in Missouri: Oil & Gas Journal, Sept. 27, 1982, p. 222-234.

Eckelmann, W.R., DeWitt, J.G. and Fisher, W.L., 1975, Prediction of fluvial-deltaic reservoir geometry, Prudhoe Bay Alaska, in, Ninth World Petroelum Congress Proceedings, v. 2, Geology: Applied Sciences Pub. Ltd., London, p. 223-227.

Eisenstatt, P., 1960, Little Creek field, Lincoln and Pike counties, Mississippi: Gulf Coast Geol. Socs. Trans., v. X, p. 206-213.

Ethridge, F.G. and Schumm, S.A., 1978, Reconstructing paleochannel morphologic and flow characteristics: methodology, limitations and assessments, in, Miall, A.D. (ed.), Fluvial Sedimentology: Canadian Soc. Petrol. Geol., Memoir 5, p. 703-721.

Ethridge, F.G. and Wescott, W.A., 1984, Tectonic setting, recognition and hydrocarbon reservoir potential of fan-delta deposits, <u>in</u>, Koster, E.H. and Steel, J.J. (eds.) Sedimentology of gravels and conglomerates: Can. Soc. Petrol. Geol. Memoir 10, p. 217-235.

Galloway, W.E., Hobday, D.K., and Magara, K., 1982, Frio Formation of the Texas Gulf Coastal plain-depositional systems, structural framework and hydrocarbon origin, migration, distribution and exploration potential: Bur. Econ. Geol. Univ., Texas, Austin, Rept. Invest. No. 122.

Galloway, W.E. and Hobday, D.K., 1983, Terrigenous clastic depositional systems, applications to petroleum, coal, and uranium exploration: Springer-Verlag, N.Y., 423 p.

Harms, J.C., 1966, Stratigraphic traps in a valley fill western Nebraska: Amer. Assoc. Petrol. Geol. Bull., v. 50, p. 2119-2149.

Harms, J.C., Tackenberg,, P., Pickles, E. and Pollock, R.E., 1981, The Brae oilfield area, in, Illing, L.W. and Hobson, g.D. (eds.), Petroleum geology of the continental shelf of northwest Europe: Heyden, p. 352-357.

Harms, J.C. and McMichael, W.J., 1982, Sedimentology of the Brae Oilfield area: Jour. Petrol. Geol., v. 5, p. 437-439.

Hayes, J.B., Harms, J.C. and Wilson, T., 1976, Contrasts between braided stream and neadering stream deposits, Beluga and Sterling Formations (Tertiary), Cook Inlet, Alaska: Alaska Geol. Soc., p. J-1 to J-27.

Jamison, H.C., Brockett, L.D. and McIntosh, R.A., 1980, Prudhoe Bay - A ten-year perspective, in, Halbouty, M.T., Giant oil and gas fields of the decade 1968-1978: Amer. Assoc. Petrol. Geol. Mem. 30, p. 289-314.

Leckie, D.A. and Walker, R.G., 1982, Storm- and tide-dominated shorelines in Cretaceous Moosebar-Lower Gates interval-outcrop equivalent of Deep Basin gas trap in western Canada: Amer. Assoc. Petrol. Geol. Bull., v. 66, p. 138-157.

Lorenz, J.C., Heinze, D.M., Clark, J.A. and Searls, C.A., 1985, Determination of widths of meander belt sandstone reservoirs from vertical downhole data, Mesaverde Group, Piceance Creek Basin, Colorado: Amer. Assoc. Petrol. Geol., Bull., v. 69, p. 710-72 .

Maughan, D.M., Mebberson, A.J. and Morton, D.J., 1981, The definition and development of the Mackerel field, Gippsland basin: Oil and Gas Journal, June, 1981, p. 175-180.

Maughan, E.K., 1984, Paleogeographic setting of Pennsylvanian Tyler Formation and relation to underlying Mississippian Rocks in Montana and North Dakota: Amer. Assoc. Petrol. Geol. Bull., v. 68, p. 178-195.

McGowen, J.H. and Bloch, S., 1985, Depositional facies, diagenesis and reservoir quality of Ivishak Sandstone (Sadlerochit Group), Prudhoe Bay field (abs): Amer. Assoc. Petrol. Geol. Bull., v. 69, p. 286.

Meister, E.M. and Aurich, N., 1972, Geologic outline and oil fields of Sergipe Basin, Brazil: Amer. Assoc. Petrol. Geol. Bull., v. 56, p. 514.

Melvin, J. and Knight, A.S., 1984, Lithofacies, diagenesis and porosity of the Ivishak Formation, Prudhoe Bay area, Alaska, in, McDonald, D.A. and Surdam, R.C. (eds.), clastic diagenesis: Amer. Assoc. Petrol. Geol. Memoir 37, p. 347-365.

Miall, A.D., 1978, Lithofacies types and vertical profile models in braided rivers: a summary, in, Miall, A.D. (ed.), Fluvial Sedimentology: Canadian Soc. Petrol. Geol. Memoir 5, p. 597-604.

Morgridge, D.L. and Smith, W.B., Jr., 1972, Geology and discovery of Prudhoe Bay field, eastern slope, Alaska: AAPG Memoir 16, p. 489-501.

Ojeda, H.A.O., 1982, Structural framework, stratigraphy and evolution of Brazilian marginal basin: Amer. Assoc. Petrol. Geol. Bull., v. 66, p. 732-749.

Orchard, D.M. and Kidwell, M.R., 1984, Morrowan stratigraphy, depositional systems, hydrocarbon accumulation in Sorrento field: Oil and Gas Journal, Sept. 3, 1984, p. 102-107.

Slack, P.B., 1981, Paleotectonics and hydrocarbon accumulation, Powder River Basin, Wyoming: Amer. Assoc. Petrol. Geol. Bull., v. 65, p. 730-743.

Sonnenberg, S.A., 1985, Tectonics and sedimentation model for Morrow Sandstone deposition, Sorrento field area, Denver Basin Colorado (abs.): Amer. Assoc. Petrol. Geol. Bull., v. 69, p. 867

Sneider, R.M., Richardson, F.M., Paynter, D.D., Eddy, R.E., and Wyant, I.A., 1977, Predicting reservoir rock geometry and continuity in Pennsylvanian reservoirs, Elk City field: Jour. Petrol. Tech., v. 29, p. 851-866.

Stone, W.D., 1972, Stratigraphy and exploration of the Lower Cretaceous Muddy Formation, Northern Powder River Basin, Wyoming and Montana: Mtn. Geologist, v. 9, p. 355-378.

Swanson, D.C., 1976, Meandering stream deposits: Cygnet Group, Inc., 30 p.

Swanson, D.C., 1979, Deltaic deposits in the Pennsylvanian Upper Morrow Formation of the Anadarko Basin, in, Hyne, N.J. (ed.), Pennsylvanian sandstones f the mid-continent: Tulsa Geol. Soc. Sp. Pub. No. 1, p. 115-168.

Tai Wai Ng, D., 1979, Subsurface study of Atoka (Lower Pennsylvanian) clastic rocks in part of Jack, Palo Pinto and Wise Counties, north-central Texas: Amer. Assoc. Petrol. Geol. Bull., v. 63, p. 50-66.

Weimer, R.J., 1983, relation of unconformities, tectonics and sea level changes, Cretaceous of the Denver Basin and adjacent areas, in, Renolds, M.W. and Dolly, E.D. (eds.), Mesozoic Paleogeography of West-Central United States: Rocky Mountain Section - SEPM, p. 359-376.

Weimer, R.J., Emme, J.J., Farmer, C.L., L.O. Anna, T.L. Davis, and R.L. Kidney, 1982, Tectonic influences on sedimentation, early Cretaceous, east flank Powder River Basin, Wyoming and South Dakota: Colorado School of Mines Quarterly, v. 77, no. 4, 61 p.

Wescott, W.A. and Ethridge, F.G., 1980, Sedimentology and tectonic setting of fan deltas-Yallaks fan delta, Southeast Jamaica: Amer. Jour. Petrol. Geol. Bull., v. 64, p. 374-399.

CHAPTER 10

OIL- AND GAS-BEARING UPPER CRETACEOUS AND PALEOGENE FLUVIAL ROCKS IN CENTRAL AND NORTHEAST UTAH

Thomas D. Fouch

U.S. Geological Survey

Golden, Colorado

INTRODUCTION

Rocks formed from mineralogically complex sediments deposited in fluvial and fluvial-lacustrine depositional settings are known from much of the world. However, relatively few of the rocks have been successfully explored for oil and gas because of the failure to adequately understand the complex lenticular reservoir units that generally characterize continental rocks. It has become increasingly obvious with new studies of hydrocarbon-bearing nonmarine rocks that in the absence of an understanding of the units sedimentologic and mineralogic rock properties, correlation of beds, and identification of genetic units and depositional facies using drill-hole logs may be inaccurate. In addition, reservoir properties of mineralogically complex rocks calculated from logs are frequently misinterpreted, and potential reservoirs may be damaged during drilling, stimulation, or completion activitites.

Abrupt changes in rock type and bedform geometry cause lenticular shaped reservoir units; variable quality within the reservoir is largely due to post-depositional mineral alteration, and induced irregularities of bedding surfaces. Oil and gas accumulations in fluvial rocks commonly occur in a complex of lens-shaped reservoirs, each with a separate hydrocarbon/water contact. Moreover, the pattern of reservoirs may be arranged in such a manner that within a sandstone unit that is continuous between drill holes, water-bearing rock occurs down structural dip from a gas/oil-bearing reservoir. Reservoirs are frequently separated by a permeability barrier of densely cemented sandstone in which the distribution of cement is at least partially controlled by sedimentary features.

Purpose

The purpose of this paper is four-fold: 1) to provide a brief description of the sedimentologic and paleontologic features found in Tertiary and Cretaceous fluvial rocks in central and northeast Utah, 2) to relate these data to depositional regimes, 3) to provide a geologic frame of reference for relating fluvial rocks to the exploration for hydrocarbons in the Uinta Basin and 4) to provide a case history (and limited model) for comparison to other hydrocarbon-bearing fluvial systems that are the focus of exploration efforts.

Nature of Reservoir Rocks

Upper Cretaceous and lower Tertiary naturally fractured, mineralogically complex, lenticular sandstones constitute the principal reservoir beds for both oil and gas in central and northeast Utah exclusive of the Paradox basin (fig. 1). The depositional and diagenetic histories of the strata are similar to those of many of the gas-bearing fluvial rocks in other areas of the Rocky Mountains. Study of these rocks offers insight into the role of sedimentology and mineralogy in the formation of hydrocarbon pools in fluvial rocks.

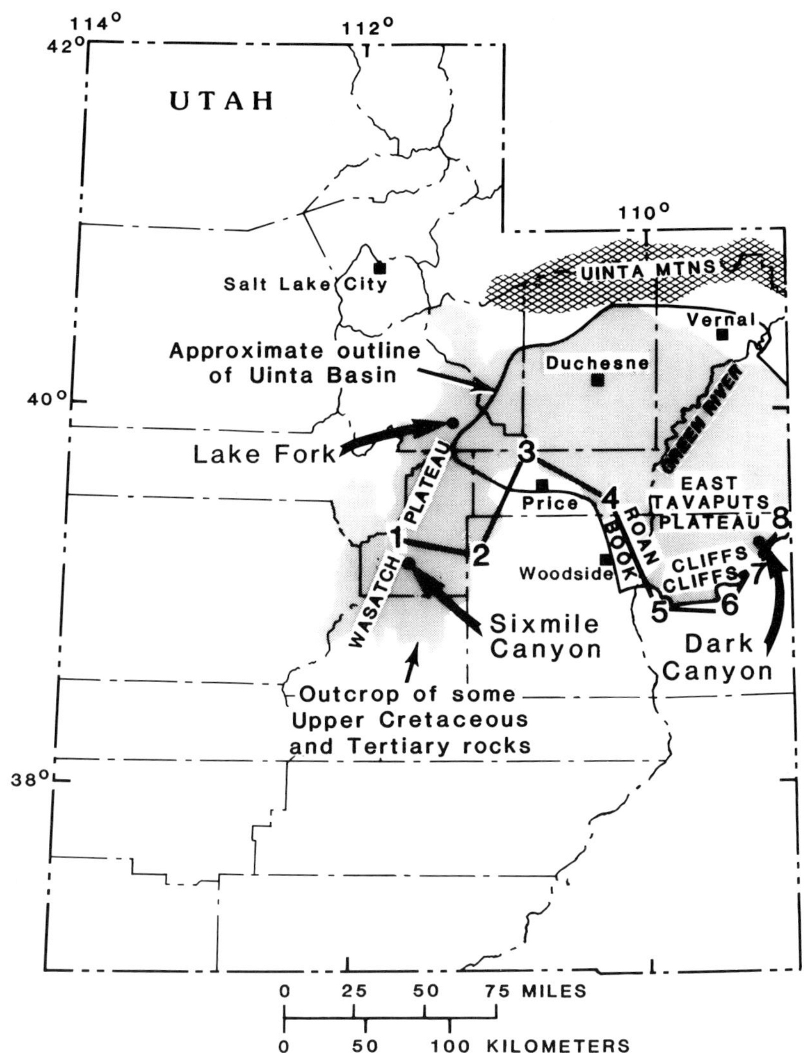

FIGURE 1.--Map of central and northeast Utah where Tertiary and Upper Cretaceous fluvial reservoir rocks contain much of the region's oil, gas, coal, and bituminous sandstones. Line of section (1-8) for sections for figure 2 are shown.

Fouch (1975), Fouch and Dean (1982), Koesemadinata (1970), and Pitman et al (1982) have indicated that rocks of Tertiary age containing the principal reservoirs for oil and associated gas fields in the region formed from sediment deposited in or near streams at the fluctuating margin of ancient Lake Uinta. Keighin (1979), Keighin and Fouch (1981), Knutson and Hodges (1981), and Pitman, et al (in press), indicate that Upper Cretaceous lenticular reservoir beds in the area are similarly of a fluvial origin. Hydrocarbons are commonly recovered from fractures and secondary pores developed from dissolution of carbonate mineral cements (Keigin and Fouch, 1981; Pitman, et al; 1982, in press).

Currently productive sandstone units in northeast Utah are characterized by highly variable but generally low values of matrix porosity and permeability (does not include fracture porosity and permeability). The Federal Energy Regulatory Commission Order 99 (1980) defined a tight reservoir as one whose in situ permeability throughout the "pay" or gas-producing section averages 0.1 md or less to gas. Reservoir rocks whose values of matrix porosity and permeabilty are low but highly variable are commonly referred to as unconventional. Data from Lucas and Drexler (1976), The National Petroleum Council (1980), Keighin and Fouch (1981), Pitman et al (1982), and Lewin and Associates (1978) indicate that matrix porosity values of productive rocks (measured in core plugs at surface conditions) range from 1 to 16 percent and matrix permeability values from several millidarcies to a few microdarcies. Boardman and Knutson (1980) studied Cretaceous and Tertiary units in six gas wells in the eastern part of the Uinta Basin. They calculated apparent matrix permeabilities of gas reservoirs using the final slope of Horner plots determined by the method of Mathews and Russell (1967). The calculated values range from 0.009 to 0.052 md and average 0.025 md. Natural open fractures provide the principal avenues of permeability among the sandstones.

More than 500 million barrels of recoverable oil have been discovered in Paleocene and Eocene rocks in the area. In addition, the National Petroleum Council (NPC) estimates that Tertiary and Upper

Cretaceous unconventional tight reservoir rocks in the Uinta Basin contain greater than 15 trillion cubic feet of recoverable gas. The NPC data indicate that most of the recovered gas will be produced from lenticular shaped reservoir units. The NPC estimates can be considered to be conservative for they were made for rocks in parts of the basin currently being explored (generally drilling depths of 10,000 ft or less. They did not include Cretaceous and lowermost Tertiary rocks in the central and western parts of the basin where the strata remain sparsley drilled and virtually unpenetrated below 17,000 ft. The magnitude of hydrocarbon resources in fluvial rocks is indeed impressive.

GEOLOGIC SETTING

Fluvial and fluvial-lacustrine rocks of early Tertiary and Late Cretaceous age in central and northeast Utah were formed from sediments deposited in response to Sevier and Laramide tectonic events. The rocks represent fluvial deposition in and adjacent to a Late Cretaceous foreland structural and depositional basin that evolved to an internally drained depositional system in Maestrichtian and Paleogene time. Figures 2 and 3 illustrate stratigraphic nomenclature and the distribution of major rock groups in time and space along a transect that extends eastward from central to northeast Utah. Many of the rocks yield hydrocarbons from the subsurface of northeast Utah and nearby areas of the Piceance basin of northwest Colorado.

Upper Cretaceous Rocks

Depositional patterns of Upper Cretaceous and some lower Tertiary

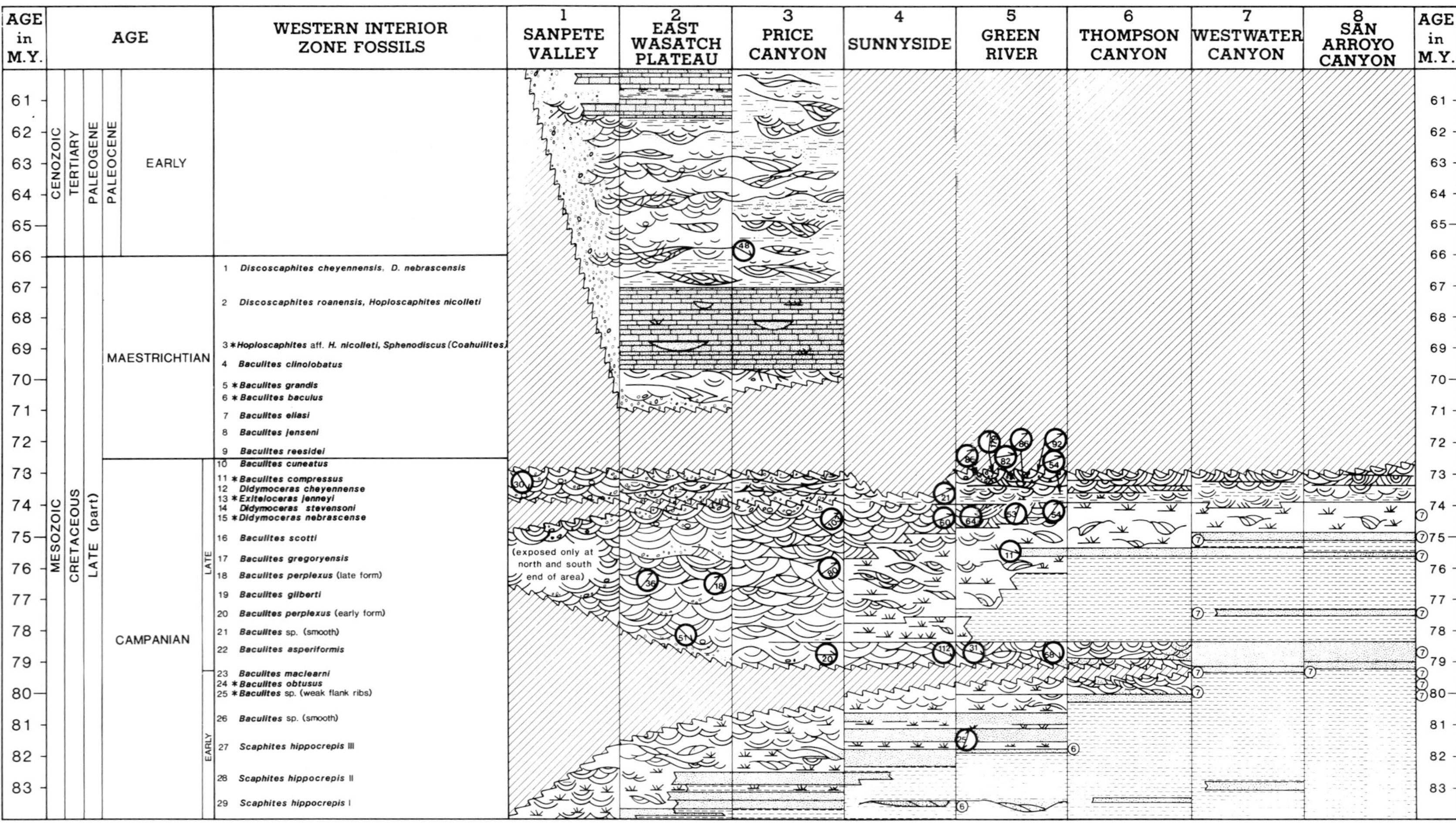

FIGURE 2.--Time-stratigraphic chart for some Upper Cretaceous and Late Paleocene rocks from central to northeast Utah. Line of section shown on figure 1. From Fouch et al (1983).

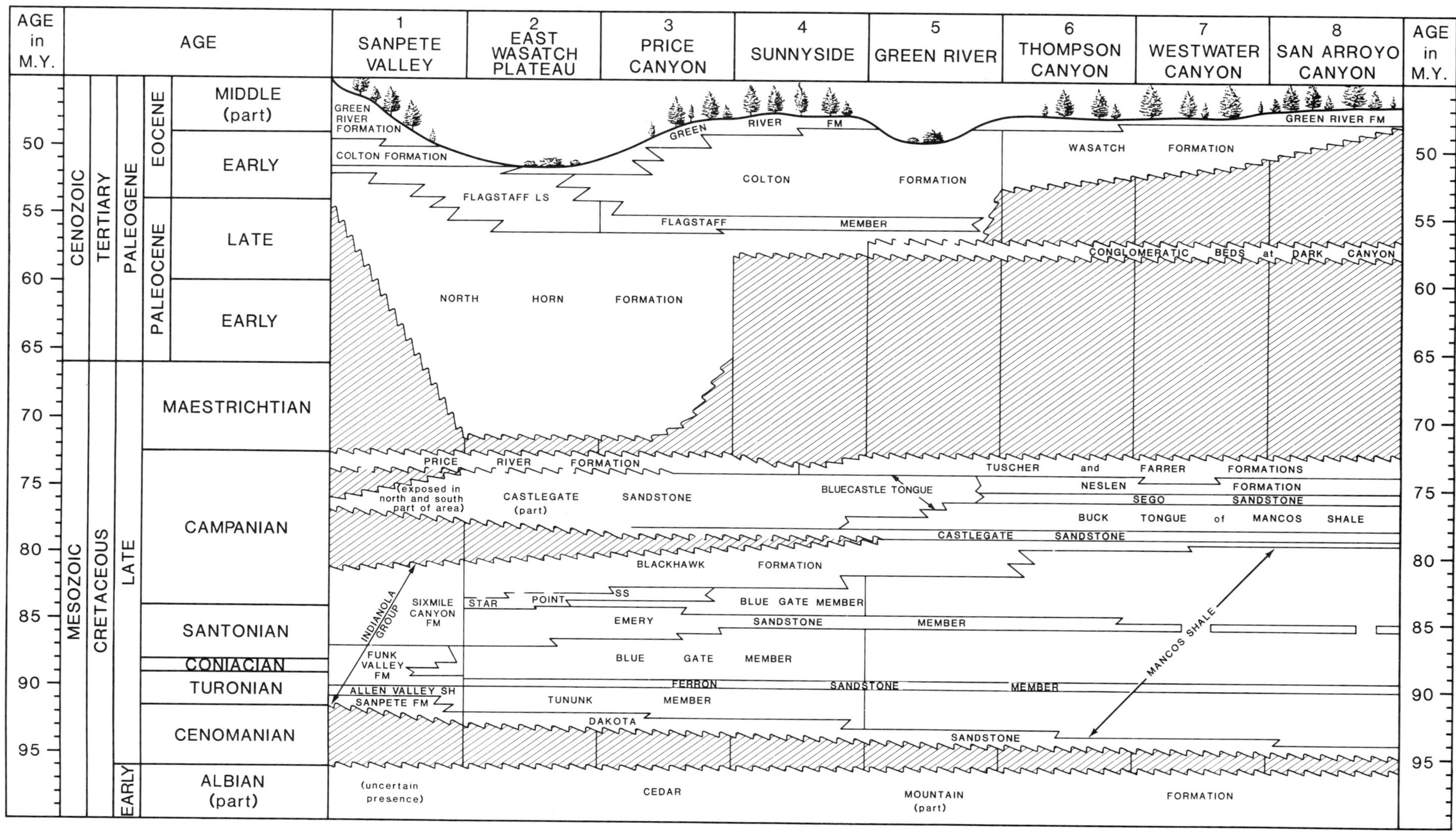

FIGURE 3.--Time-stratigraphic chart illustrating stratigraphic nomenclature and correlation of major Albian to middle Eocene units from the Sanpete valley of central Utah to the Book Cliffs of eastern Utah. Line of section shown on figure 1. Vertical line through strata indicates a change in stratigraphic nomenclature. Taken from Fouch et al (1983).

rocks in central and northeast Utah were controlled by thrust-fold events in eastern and southern Nevada, and western and central Utah. Figure 4 is a cross section that illustrates lithologies of Upper Cretaceous through early Eocene beds from Price Canyon in central Utah to the eastern Book Cliffs. The lithologic groups can be traced to the subsurface of eastern Utah and western Colorado where they form a depositional system that contains indigenous source, reservoir, and trap rocks for both oil and gas.

Upper Cretaceous rocks can be assigned to five major lithofacies which record the transition from sedimentation in a nonmarine to marine foreland depositional basin to sedimentation in an internally drained hydrologic basin. The descriptions of subdivisions of lithofacies for Upper Cretaceous rocks are taken principally from Fouch et al (1983). The first four lithofacies are described in the order of their proximal (west) to distal (east) depositional relations. The fifth facies is composed of rocks formed from sediments deposited in an internally drained basin.

I. Sheet Conglomerate Lithofacies

The sheet conglomerate lithofacies is located principally in the western part of the area generally west of the Wasatch Plateau and in proximity to the Sevier orogenic highland. Some conglomeratic beds are developed along the north and west margin of the present day Uinta Basin. Units of the conglomerate facies can be characterized as follows:

a. units consist of abundant conglomerate with lesser amountsof pebbly sandstone, debris-rich siltstone, and red to gray noncalcareous claystone;

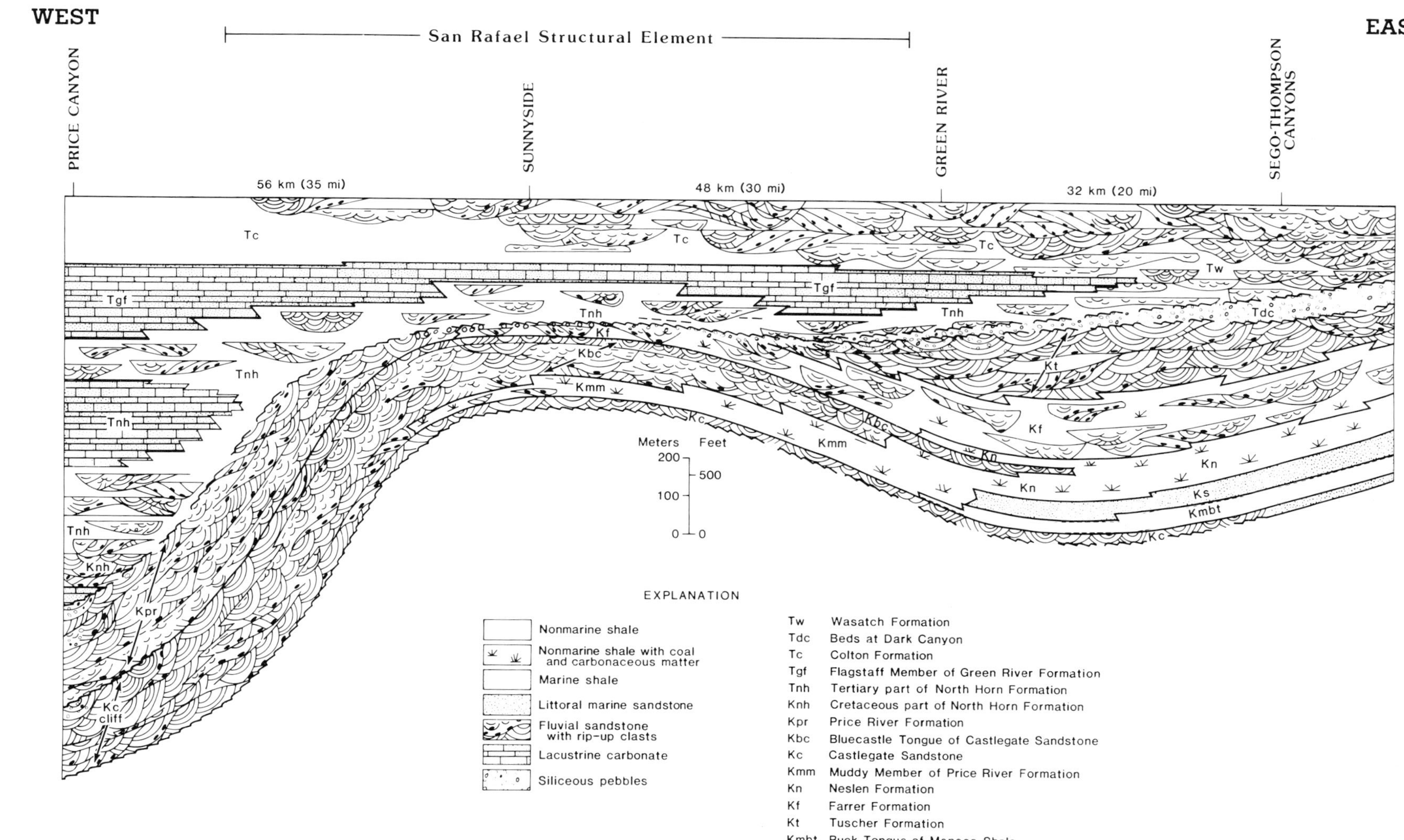

FIGURE 4.--Stratigraphic cross section of some Campanian through Paleocene rocks extending eastward from Price Canyon to Sego-Thompson Canyons showing lithofacies and interpreted depositional environments. See figure 1 for line of section. Taken from Fouch et al (1983).

b. conglomerate units are generally clast supported and internally structureless;

c. conglomeratic units form near flat and laterally extensive units that range in thickness from thin beds to near 20 ft;

d. near their distal margins, some conglomeratic beds contain relatively small and isolated channels; some conglomerates are imbricated and crudely graded;

e. carbonaceous matter and sparse root traces are the only indicators of indigenous life.

Rock textures, bedform geometry, and paleontology collectively indicate that the conglomeratic sheets represent fanglomerate deposits formed during episodic sheet-flooding and debris-flow events that occurred along the east margin of the Late Cretaceous lushly vegetated subtropical highland.

Rocks of the sheet conglomerate facies are not known to contain hydrocarbons in the region.

II. Scour-fill Conglomerate Lithofacies

The scour-fill conglomerate facies can be characterized by:

a. units that are composed of commonly crudely imbricated and graded channel-form conglomerates and conglomeratic sandstones;

b. conglomeratic units that form laterally extensive near flat sheets and wedges composed of coalesced channel-form beds. Each channel-fill conglomerate is in sharp erosional contact with underlying units;

d. fossils that are restricted to carbonaceous matter, rare root(?), limb and leaf traces and sparse bone fragments.

The abundance and ubiquitous presence of crudely imbricated and graded, coalesced channel-fill cycles, and the regional setting collectively suggest the beds originated from episodic periods of flow in braided-channels on an alluvial fan or braidplain. Each sheet may represent the development of an individual fan lobe.

III. Pebbly Trough-crossbedded Lithofacies

The facies consists of pebbly channel-fill sandstone with well developed large-scale trough crossbeds. The beds are characterized by:

a. channels which combine to form large extensive sandstone sheets as much as 100 ft thick, some of which extend laterally for several miles. Individual sandstone sheets can frequently be correlated on borehole logs for a distance nearly equal to the size of the sheets;

b. well-developed cycles in which grain size and scale of cross stratification decrease upward from a basal scour;

c. palynomorphs and carbonaceous matter are preserved along some sandstone laminae and in thin lenses between large-scale low-angle crossbeds most commonly as carbonaceous clasts;

d. plant debris and root traces represent the principal fossil material;

e. intervals between sandstone sheets are characterized by fine-grained sandstone and siltstone scour-fill beds and laminated to structureless siltstone and claystone rich in woody plant material.

The lithofacies assemblege can be interpreted to have formed from sediments deposited in relatively straight to moderately sinuous mixed-load and bed-load streams on the lower part of an alluvial braidplain. Rocks that comprise the intervals between blanketlike sandstone sheets are inferred to have formed in overbank and other flood-plain settings crossed by mixed load moderately sinuous streams.

The pebbly trough-crossdedded facies constitutes an important productive assemblage of Cretaceous rocks in eastern Utah. Reservoir units of the Tuscher Formation and the upper part of the Farrer Formation of the Mesaverde Group yield gas where penetrated well below the unconformity between Cretaceous and Tertiary rocks in the southeastern part of the Uinta Basin. Although the sandstone sheets that contain the reservoir beds can be correlated on logs from drillholes as much as one half of a mile apart, the reservoirs rocks commonly do not extend more than a few tens of feet. Pores in reservoir sandstones commonly are secondary and individual reservoirs generally are relatively small and isolated.

IV. Accretion and Scour-fill Lithofacies

The facies is characterized by:

a. channel-form sandstone units that are generally composed of laterally accreted large-scale low-angle crossbeds, and flat-bedded to visually structureless siltstone and claystone beds;

b. flat-bedded units that contain abundant terrigenous lithic and organic debris are commonly interbedded with laminae to thin beds of coal;

c. cross stratification within sandstone units is generally cyclic and of a larger scale and angle near the base of individual trough;

d. channel units that are commonly much less than 30 ft thick and less than 500 ft wide;

e. coal beds that extend laterally several tens of feet laterally;

f. lithologies that all are rich in rafted and rooted woody plant material;

g. typical fossils are plant remains and sparse nonmarine bivalves; charophtyes are rare.

The lithologic and paleontologic associations indicate the rocks were formed from sediments deposited on a poorly drained alluvial plain which extended from the toe of an alluvial fan to the marine coastline. The plain was crossed by moderately sinuous to sinuous streams, and dotted with well-developed swamps and lakes.

Lenticular sandstone reservoirs of the middle and lower part of the Farrer Formation and of the coal-bearing Neslen Formation yield gas in the eastern part of the Uinta Basin (see Keighin and Fouch, 1981). The reservoir rocks occur in sandstone units that rarely correlate between closely spaced drillholes.

V. Calcareous Claystone and Carbonate Rock Lithofacies

Rocks of this facies are characterized by:

a. thinly laminated to structureless thin- and thick-bedded calcareous rocks;

b. beds that are generally near flat;

c. shrinkage cracks, rootlet traces, vugs, and lithoclasts that are all locally abundant;

d. virtually all beds yield calcareous microfossils of which charophytes are most abundant; ostracodes, woody plant material, palynomorphs and bone fragments are locally abundant. Dinoflagellates are sparse.

Rocks of this facies formed from sediments that were deposited in an internally-drained hydrographic basin. Paleontologic data indicate the lake(s) was large and contained abundant calcareous algae that lived in relatively shallow, warm, slow-moving alkaline water.

Rocks of this facies have been recognized only in Maestrichtian strata in northcentral Utah (fig. 2). Lake beds commonly contain lipid-rich organic matter which can be transformed into oil upon thermochemical maturation.

Fluvial rocks that intertongue with units of the lacustrine facies may form the basal part of the oil-bearing section at the giant Altamont-Bluebell field in the northcentral part of the Uinta Basin (Fouch, 1981).

Paleogene Rocks

The Paleogene rocks of northeast Utah are particularly noteworthy, for in addition to containing fluvial rock reservoirs, they contain lipid-rich lacustrine beds that are the sources of oil in the large oil fields of the Uinta Basin.

Lithofacies groups are complexly interbedded in the basin because the nature of the hydrologic basin continually changed during early Eocene time and the lake's shoreline frequently fluctuated over large areas. An additional changing feature of Lake Uinta that affected the distribution of sediment, and floral and fauna types, was the intermittent extreme alkalinity and salinity of water in the central part of the lake. As the level of the lake fluctuated, dense saline lake water locally extended some distance into subaerial deltaic settings and locally affected the capacity of the subaerial streams to transport sediment into the lake.

Ryder et al (1976) and Fouch (1975) subdivided the lithofacies groups of Paleogene rocks in the Uinta Basin and some adjacent areas and described and illustrated their concept of the model that best represents their interpretation of the depositional regime in effect in early Eocene time in this part of the central Rocky Mountains (figure 5). The model was formulated to accomodate both surface and subsurface data, and to relate paleontologic,

organic geochemical, sedimentologic, and engineering data to exploration for oil and gas. Strata below the middle marker (see fig. 5) have been divided into three major intertonguing depositional facies: 1) alluvial, 2) marginal lacustrine, and 3) open lacustrine.

I. Alluvial Depositional Facies

The alluvial facies is composed primarily of argillaceous sandstone, conglomerate, siltstone and claystone; sparse carbonate units are locally developed. The facies can be subdivided in to several subfacies whose characteristics are given below.

a. Alluvial fan facies: (fig. 6)

1. developed adjacent to mountain fronts;
2. is composed of conglomerate, sandstone, and siltstone;
3. contains thick, structureless to crudely stratified near flat sheets and wedges; some extend to near one half of a mile;
4. formed from braided channel deposits on a braidplain and from sparse debris flows. Each sheet may be a fan lobe.

Alluvial fan conglomeratic rocks are exposed in Tertiary strata at the western end of the Uinta Basin and sheetlike conglomeratic rock complexes are developed along the basin's southeast flank where the conglomeratic units were referred to as the beds at Dark Canyon by Fouch and Cashion (1979). These siliceous quartzitic pebble conglomerates can be traced from surface exposures to the subsurface of the basin where

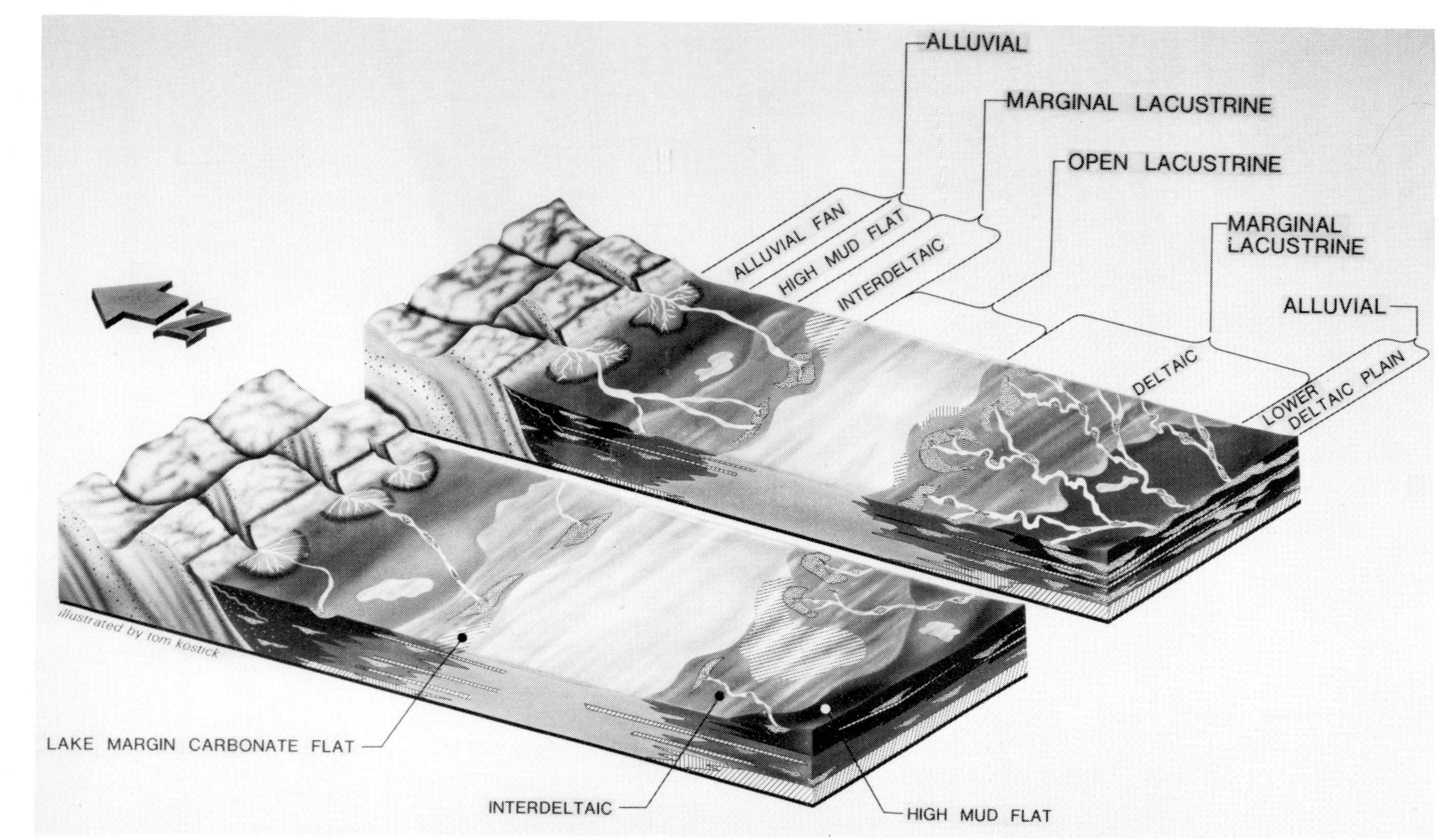

FIGURE 5.--Block diagram illustrating the distribution of interpreted open-lacustrine, marginal-lacustrine, and alluvial depositional environments of the western portion of Lake Uinta, Utah, as it existed during the early Eocene. Diagonally striped pattern is grain-supported carbonate rock, heavy dots, sandstone; dark-gray tone, red claystone; medium-gray tone, gray and green claystone; light-gray tone, dark-gray and brown claystone. Width of Lake Uinta in the diagram is approximately 40 km. Vertical exaggeration is between 15 and 20. Modified from Ryder et al (1976).

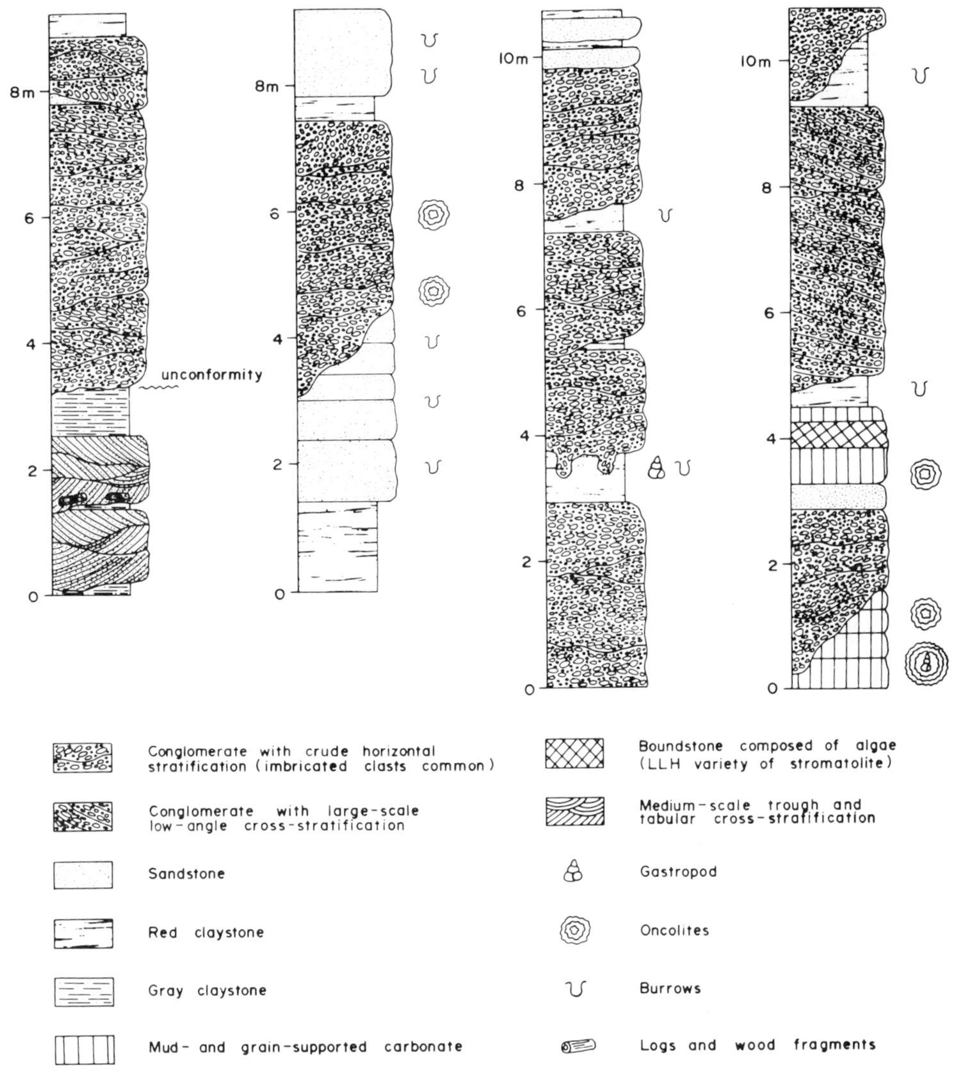

FIGURE 6.--Rock sequences that originated in the alluvial-fan environment. Taken from Ryder et al (1976).

they are identified on drillhole logs. The beds are reservoirs for oil at Fence Canyon field (Petersen, 1973) in the southeast part of the basin and bear gas in the general region of Natural Buttes (Keighin and Fouch, 1981). Uranium prospect pits are developed in Dark Canyon beds at the base of the Roan Cliffs, east of the Green River in central Utah. Although beds of this depositional facies are poorly exposed in the Wasatch Mountains and at the western end of the basin they may form fan deltas intercalated with shallow lacustrine carbonate units. The

extent of the facies in the subsurface of the northern part of the basin is poorly known. However, the facies extends from the basin boundary fault (known only from the subsurface) illustrated by Ryder et al (1976) to near the edge of, and perhaps locally into, Lake Uinta in the northwest part of the basin (fig. 7).
Conglomeratic fan deltas might be present in this area.

b. Lower deltaic plain facies (fig. 8):

1. thin to medium flat-bedded red claystone beds with well-developed mudcracks and local bioturbation;
2. claystones commonly intercalated with laterally continuous fine-grained sandstones exhibiting small-scale crossbeds (local climbing ripples), and structureless units;
3. composite channel-form units may be dominant. Near the edge of the lake, many channels contain well-developed large-scale low-angle crossbeds that have the appearance of deltaic foreset beds or accretion beds. The large scale crossbeds contain their coarsest grains near the base of the channel scour.
4. Internal scoured surfaces are locally well developed in broad flat-based channels.

Rocks formed from sediment deposited in lower deltaic-plain settings at the margin of Lake Uinta are locally intercalated with lacustrine units formed in periods of extensive expansion of the lake. In such cases, the lacustrine and alluvial facies become difficult to separate in the subsurface and in areas of poor surface exposures; the beds are frequently arbitrarily grouped with beds of the marginal lacustrine facies. However, it is useful to separate those fluvial rocks formed in settings well removed from the lake from those deposited

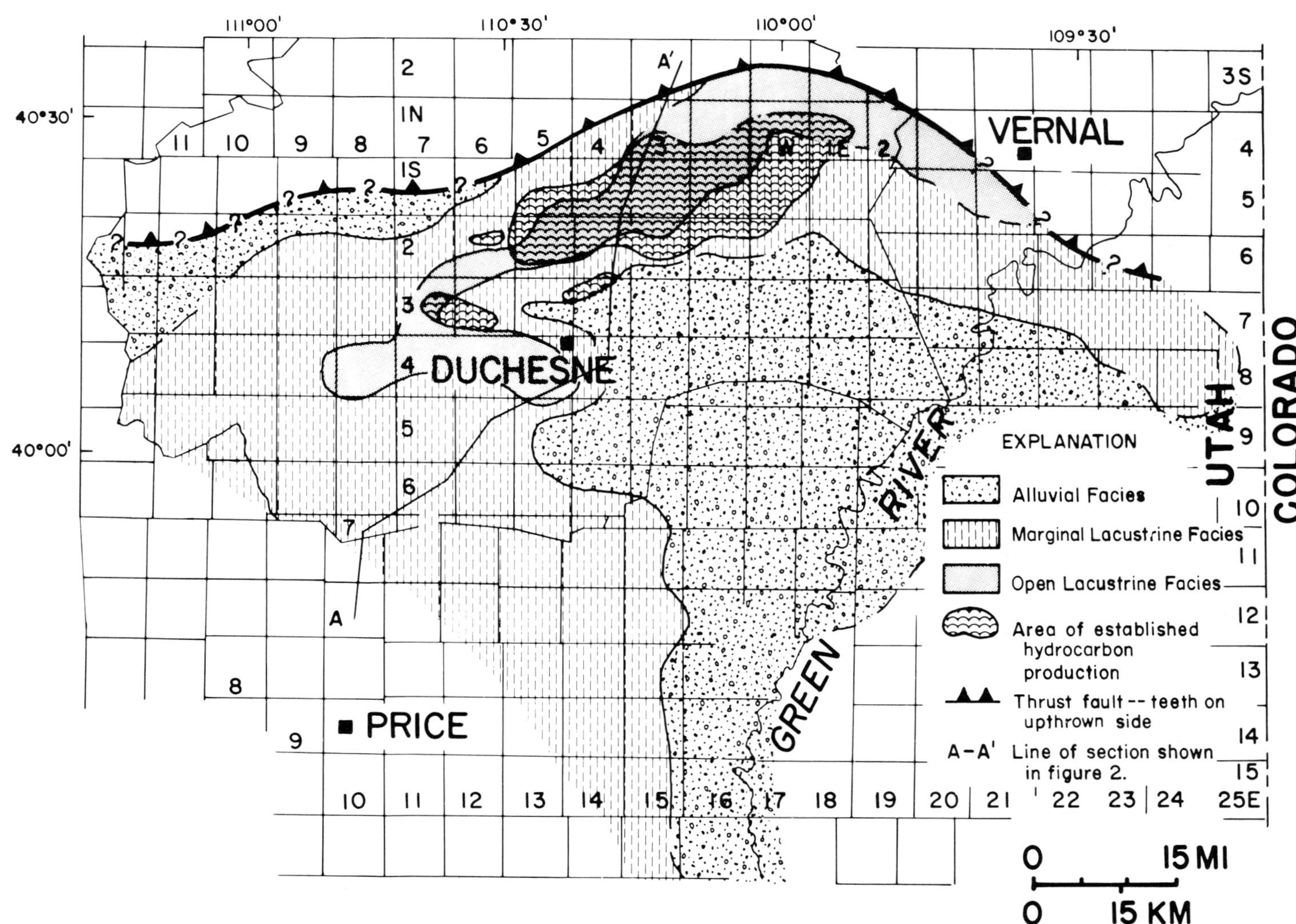

FIGURE 7.--Generalized lithofacies map showing areas of established production in 1975 from a zone of rocks consisting of beds adjacent and laterally equivalent to the lower marker of the Flagstaff Member of the Green River Formation. Taken from Fouch (1975).

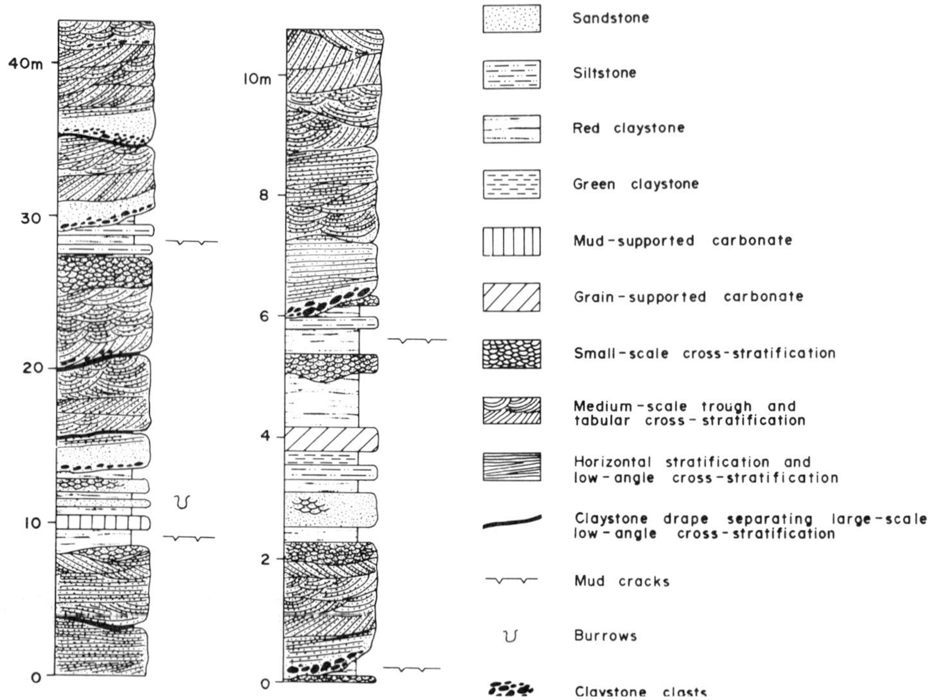

FIGURE 8.--Rock sequences that originated in the lower deltaic-plain environment. Taken from Ryder et al (1976).

at or near the fluctuating margin of the lake. Fluvial rocks deposited on the lower deltaic plain in settings well removed from the lake commonly are reservoirs for only small amounts of oil in Utah whereas channel sandstones formed at the fluctuating margin of the lake are the principal reservoirs for oil in the Uinta Basin. However, gas is recovered from channels deposited in areas that were infrequently covered by lake water.

c. High mudflat depositional facies:

The facies:

1. is laterally adjacent to lower deltaic plain;
2. contains red claystone and minor isolated channel-fill units;
3. was subaerially exposed much of the time with isolated ponds that commonly contained saline water. Some ponds contained sulfate salts which locally cemented upper Paleocene and lower Eocene

carbonate grainstones and sandstones;

4. contains common mudcracks, rooted rocks.

Rocks formed from sediment deposited on an alluvial high mud flat adjacent to the Paleogene lakes of Utah generally constitute a trapping facies for hydrocarbons. Oil and gas moving up depositional dip through interconnected pores or open fractures in marginal-lacustrine rocks at the edge of the lake are trapped by the relatively nonporous and impermeable beds of the claystone-dominated high mudflat. Beds of this facies are relatively ductile and contain relatively few interbeds with major contrasts in brittleness. For this reason, natural fractures are relatively sparse when compared to units of the sandstone and carbonate-rich facies.

II. Marginal Lacustrine Depositional Facies

The marginal lacustrine facies is composed of sandstone, claystone, and mixed grain- and mud-supported carbonate units. Rocks formed near the edge of the lake or during freshwater lake phases are fossiliferous. The predominant claystone colors range from light- and medium-gray to gray-green. Rocks present in the facies are interpreted to have formed from sediments deposited in deltaic, interdeltaic, and lake-margin carbonate-flat environments. The facies represents both lacustrine and subaerial depositional settings at the fluctuating margin of Eocene and Paleocene Lake Uinta.

a. Deltaic (fig. 9)

Rocks of the facies are characterized by:

1. individual and composite channel deposits that are dominant;

2. channel-fill cycles that formed in streams at the fluctuating margin of the lake in areas frequently submerged beneath the lake water;

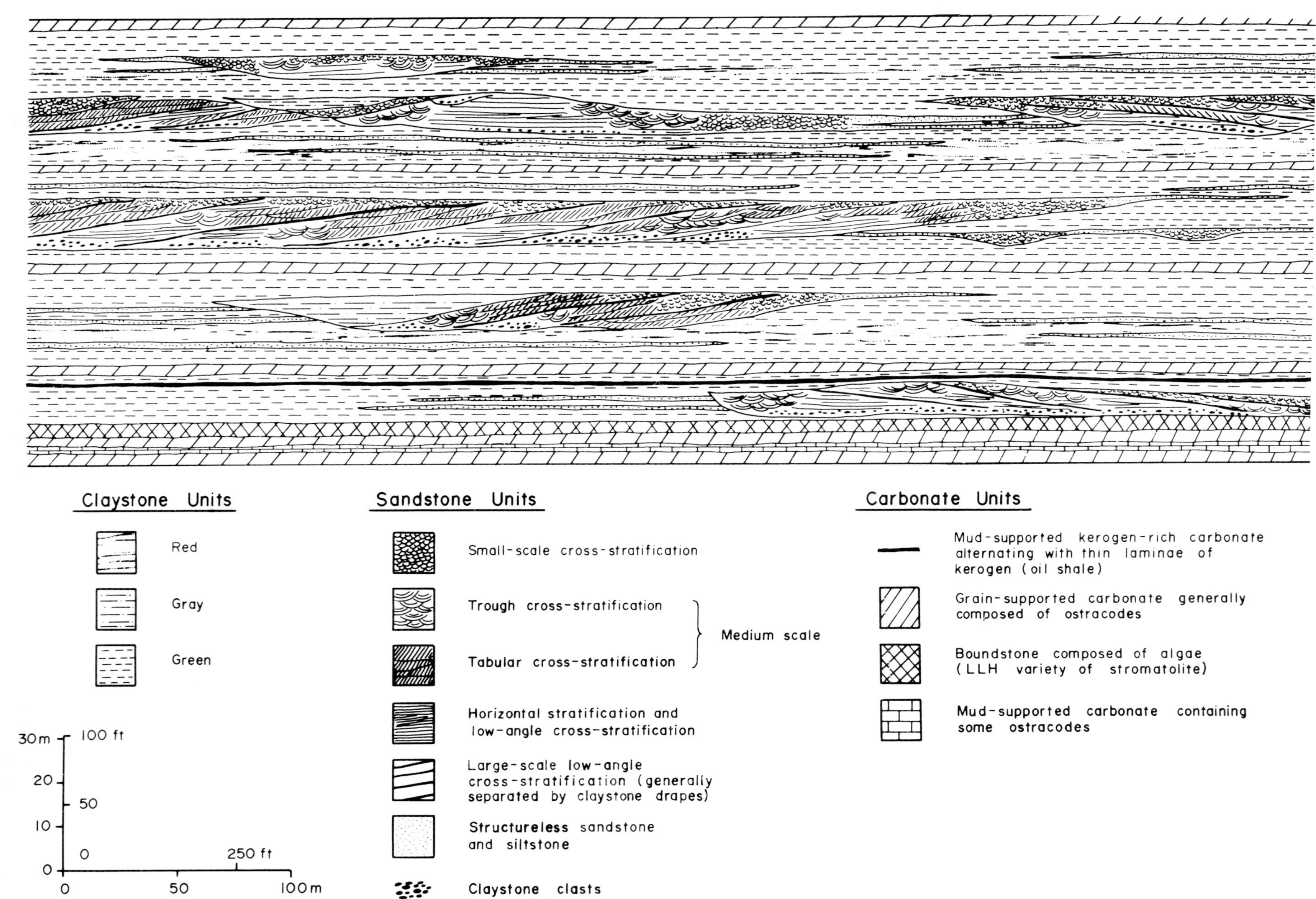

Claystone Units
Red
Gray
Green
30m
20
10
0
100 ft
50
0
250 ft
0
50
100m
Sandstone Units
Small-scale cross-stratification
Trough cross-stratification
Tabular cross-stratification
Medium scale
Horizontal stratification and low-angle cross-stratification
Large-scale low-angle cross-stratification (generally separated by claystone drapes)
Structureless sandstone and siltstone
Claystone clasts
Carbonate Units
Mud-supported kerogen-rich carbonate alternating with thin laminae of kerogen (oil shale)
Grain-supported carbonate generally composed of ostracodes
Boundstone composed of algae (LLH variety of stromatolite)
Mud-supported carbonate containing some ostracodes

FIGURE 9.--Distribution of rocks that originated in the deltaic environment: channel-form sandstone beds, delta distributary channels; thin sandstone and siltstone beds, delta-front and overbank deposits undifferentiated; claystone beds, interdistributary nearshore lacustrine deposits; kerogen-rich bed, open-lacustrine deposit; other carbonate beds, lake-margin carbonate-flat deposits. Taken from Ryder et al (1976).

3. sinuousity of streams that varied as lake level, stream water velocity, and sediment load fluctuated;
4. contrasts in water salinity between that of the saline-stratified lake and relatively fresh streams that influenced distribution of grain sizes and bedforms;
5. energy of lake water was generally low as relatively few stream deposits were reworked to form thick or extensive lacustrine bars.

Deltaic marginal-lacustrine rocks form the principal reservoirs for oil in the Uinta Basin. These reservoir units were most commonly formed in subaerial streams in areas frequently submerged beneath the water of Lake Uinta. Lacustrine bar deposits may locally be reservoir units (Picard and High, 1970). Marginal-lacustrine rocks are also the principal reservoir beds for several billion barrels of oil in place in the bituminous sandstone units of the basin (Campbell and Ritzma, 1979; Ritzma, 1979). The giant oil-impregnated sandstone deposits exposed at several localities along the south flank of the Uinta Basin are developed principally in lenticular units. Most lenticular reservoir units are of a fluvial origin. Porosity and permeability values in these rocks are very high (see Byrd, 1970). The bitumen occurs in sandstone and carbonate rocks with little or no cement and in pores developed within pisolites and carbonate grains coated with carbonate minerals. The porous and permeable rocks grade down structural dip into the subsurface of the Uinta Basin where depositionally and temporally equivalent beds contain oil in secondary pores. In the subsurface of most of the basin, values of porosity and permeabilty are extremely low (see INTRODUCTION). Values of porosity are greatest on the southern most exposures of the basin where near-surface dissolution of mineral cements apparently is most extensive.

b. Lake-margin carbonate-flat

The rocks of the facies are characterized by:

1. a systematic basinward gradation from sandstone to gray claystone (may include algal coal) to oncolite and (or) oolite grain-supported carbonate or stromatolitic carbonate to mud-supported kerogenous carbonate;
2. Common Shoal cycles;
3. Common molluscks, ostracode fossils, some charophytes;

Rocks formed in the carbonate-flat setting are generally brittle relative to associated lithologies. As a result, the beds are commonly fractured both on surface exposures and in the subsurface. The natural open fractures serve as the principal avenues of permeability in oil and gas filds developed in Tertiary rocks in the region.

c. Open Lacustrine Depositional Facies

The facies is characterized by:

1. calcareous claystone and mud-supported carbonate units;
2. rocks that are kerogenous;
3. structureless to flat-laminated rocks.

Open-lacustrine rocks of the Uinta Basin contain the principal source beds for the oil accumulations of the basin (Tissot et al, 1978).

TRAPS

Oil and gas pools in Upper Cretaceous and Tertiary rocks in

northeast Utah can be considered to occur in stratigraphic traps. Pores in rock reservoirs are generally secondary, a phenomenom which indicates that perhaps the traps should be called diagenetically modified stratigraphic traps. The overpressured rocks of the giant Altamont-Bluebell field yield most of the oil and gas from the western part of the Uinta Basin. Figure 10 illustrates that marginal-lacustrine strata extend from overpressured subsurface beds at Altamont to the surface along the basin's south flant. Fluvial rocks in these marginal-lacustrine strata are impregnated with oil where exposed. This occurrence of oil-bearing surface rocks that extend to subsurface over-pressured rocks in the Altamont-Bluebell field indicates that the distribution of hydrocarbons is not limited by a conventional trap or a lack of reservoir rocks. In a sense, the Altamont-Bluebell field has no conventionalt trap at its southern margin and is an accumulation that extends to the surface. The southern limits of the productive section of Altamont-Bluebell is defined only by the economic and technicalogical limits of recovering the high pour-point oil from lenticular rocks. Overpressured rocks at Altamont-Bluebell, although buried deeply, yield hydrocarbons from fractured rocks of low porosity and matrix permeability because of the abnormally high pressures. In the deep subsurface, pressures are sufficient to move oil through narrow pore throats to open fractures connected to wellbores.

Although oil and gas exploration in fluvial rocks of the Uinta Basin may prove to be most successful by simply drilling in overpressured rocks such as at Altamont-Bluebell, normally pressured fluvial beds also yield hydrocarbons. Keighin and Fouch (1981) and Pitman et al (1982, in press) found that porosity in fluvial rocks is

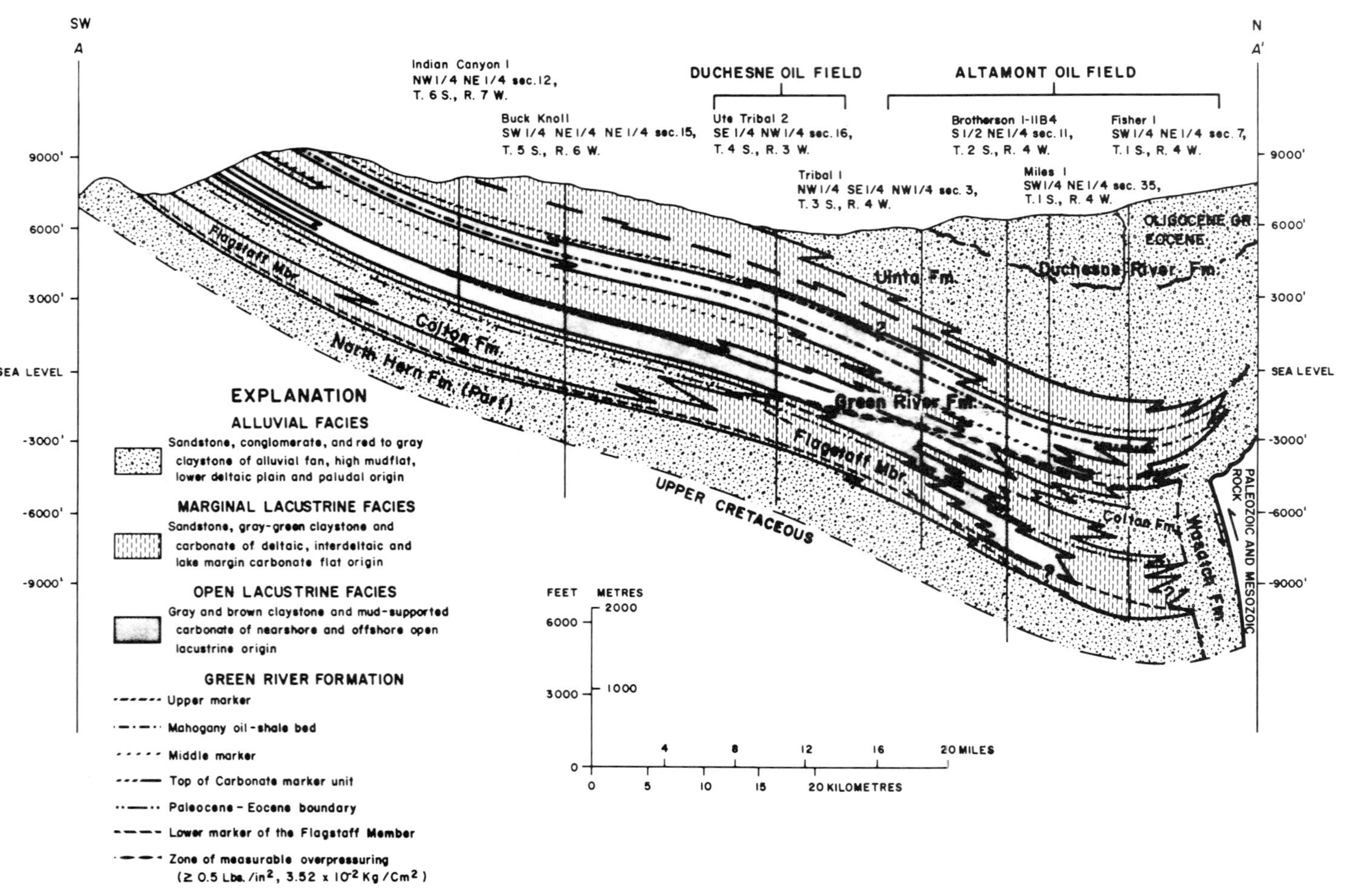

FIGURE 10.--Generalized structural-stratigraphic cross sesction from outcrops on the southwest flank of the Uinta Basin, through Duchesne and Altamont-Bluebell oil fields, to the north-central part of the basin. Uinta Formation includes saline facies and equivalent lacustrine rocks assigned to the Uinta by Dane (1954). Taken from Fouch (1975).

most commonly developed in the basal parts of channels where there has been extensive dissolution of carbonate cement. Some cored fluvial channels in the Mesaverede in the region of Natural Buttes field may contain as much as 25% carbonate mineral cement. Where amounts of carbonate cement are high, the rocks typically contain little porosity and may be fractured.

The variable distribution of mineral cements and the complex mineralogy of the detrital grains in fluvial rocks can significantly alter the response of commonly used borehole logs from that typical of mineralogically simple lithologies. Therefore, employing logs to determine porosity or the origin of genetic units in mineralogically complex rocks should be approched cautiously.

SUMMARY STATEMENT

Fluvial rocks constitute the principal reservoir units for extremely large volumes of oil and gas in northeast Utah. The hydrocarbons commonly reside in secondary inter- and intra-granular pores developed in rocks of a complex mineralogy. Similar fluvial reservoir beds have been explored in much of the world but many attempts to do so have been economic failures. Exploration and exploitation activities must employ some knowledge of the sedimentology and mineralogy of the rocks to increase the ability to detect and succesfully recovery the oil and gas.

REFERENCES CITED

Boardman, C. R., and Knutson, C. F., 1980, Reservoir characteristics in Uinta Basin gas wells, final report for the period 1 September 1978

- January 1980: U. S. Department of Enegry Report DOE/ET/11399-1, 89 p., Report is available from National Technical Information Service, U.S. Department of Commerce, Springfield, Virginia, 22161.

Byrd, William, D. III, 1970, P. R. Spring oil-impregnated sandstone deposit, Uinta and Grand Counties, Utah: Utah Geological and Mineralogical Survey Special Studies 31, 34 p.

Campbell, Jock A., and Ritzma, H. R., 1979, Geology and petroleum resources of the major oil-impregnated sandstone deposits of Utah: Utah Geological and Mineral Survey Special Studies 50, 24 p.

Dane, C. H., 1954, Stratigraphic and facies relationships of upper part of Green River formation and lower part of Uinta formation in Duchesne, Uintah, and Wasatch Counties, Utah: American Association of Petroleum geologists Bulletin, v. 38, p. 405-425.

Federal Energy Regulatory Commission, 1980, High-cost natural gas produced from tight formations: U.S. Federal Energy Regulatory Commission Order No. 99, Docket No. RM79-76, 65 p.

Fouch, T. D., 1975, Lithofacies and related hydrocarbon accumulations in Tertiary strata of the western and central Uinta Basin, Utah, in Bolyard, D. W., ed., Symposium on Deep Drilling Frontiers in the Central Rocky Mountains: Rocky Mountain Association of Geologists, p. 163-173.

______ 1981, Chart showing distribution of rock types, lithologic groups and depositional environments for some lower Tertiary and Upper Cretaceous rocks from outcrops at Willow Creek-Indian Canyon through the subsurface of the Duchesne and Altamont oil fields, southwest to north-central parts of the Uinta Basin, Utah: U.S. Geological Survey Oil and Gas Investigations Chart, OC-81, 2 sheets in color.

Fouch, T. D. and Cashion, W. B., 1979, Distribution of rock types, lithologic groups, and depositional environments for some lower Tertiary and Upper and Lower Cretaceous, and Upper and Middle Jurassic rocks in the subsurface between Altamont oil field and San Arroyo gas field, nothcentral to to northeastern Uinta Basin, Utah: U.S. Geological Survey Open-file Report 79-365, 2 sheets.

Fouch, T. D. and Dean, 1982, Lacustrine and associated clastic depositional environments, in Scholle, P. A., and Spearing, D., eds., Sandstone Depositional Environments: American Association of Petroleum Geologists Memoir 31, p. 87-114.

Fouch, T. D., Lawton, T. F., Nichols, D. J., Cashion, W. B., and Cobban, W. A., 1983, Patterns and timing of synorogenic sedimentation in Upper Cretaceous rocks of central and northeast Utah; in Reynolds, M., Dolly, E., and Spearing, D., eds., Mesozoic Paleogeography of West-central United States, Society of Eoconomic Paleontologists and Mineralogists, Rocky Mountain Section Symposium 2, p. 305-334.

Keighin, C. W., 1979, Influence of diagenetic reactions on reservoir properties of the Neslen, Farrer, and Tuscher Formations, Uinta Basin, Utah: Society of Petroleum Engineers Paper SPE 7919, presented at the 1979 SPE Symposium on Low-Permeability Gas Reservoirs, May 20-22, 1979, Denver, CO, p. 77-84.

Keighin, C. W. and Fouch, T. D., 1981, Depositional environments and diagenesis of some nonmarine Upper Cretaceous reservoir rocks, Uinta Basin, Utah: _in_ Ethridge, F. G., and Flores, R. M., eds., Recent and Ancient Nonmarine Depositional Environments: Models for Exploration, Society of Eoconomic Paleontologists and Mineralogists

Special Publication 31, p. 109-125.

Knutson, C. F., and Hodges, L. T., 1981, Development of techniques for optimizing selection and completion of western tight gas sands, comparison of core, geophysical log, and outcrop information, phase III report: U.S. Department of Energy Report DOE/BC10005-3, 54 p., available from National Information Service, U.S. Department of Commerce, Springfield Virginia, 22161.

Koesoemadinata, R. P., 1970, Stratigraphy and petroleum occurrence, Green River Formation, Redwash Field, Utah: Colorado School of Mines Quarterly; v. 65, 77 p.

Lewin and Associates, Inc., 1978, Enhanced recovery of unconventional gas, main report, v. 11, U. S. Department of Energy, 336 p.

Lucas, P. T., and Drexler, J. M., 1976, Altamont-Bluebell -- a major naturally fractured stratigraphic trap, Uinta Basin, Utah, <u>in</u> North American Oil and Gas Fields: American Association of Petroleum Geologists Memoir 24, p. 121-135 p.

Mathews, C. S., and Russell, D. G., 1967, Pressure buildup and flow tests in wells: Henry L. Doherty Series, SPE-AIME Monograph Volume I.

Meyerhoff, A. A., and Willums, J. O., 1976, Petroleum geology and industry of the People's Republic of China: United Nations ESCAP, CCOP, Technical Bulletin, v. 10, p. 103-212.

National Petroleum Council, 1980, Tight gas reservoirs - Part II, unconventional gas sources: J. F. Bookout, Chairman, Committee on Unconventional Gas Sources, 749 p.

Peterson, P. R., 1973, Fence Canyon field: Utah Geological and Mineralogical Survey Oil and Gas Field Studies No. 9, 4 p. 1 map.

Picard, M. D., and High, L. R., Jr., 1970, Sedimentology of oil-impregnated lacustrine and fluvial sandstone, P. R. Spring area southeast Uinta Basin, Utah: Utah Geological and Mineralogical Survey Special Studies 33, 32 p.

Pitman, J. K., Fouch, T. D., and Goldhaber, M. B., 1982, Evolution of some Tertiary unconventional reservoir rocks, Uinta Basin, Utah: American Association of Petroleum Geologists Bulletin, v. 66, No. 10, p. 1581-1596.

Pitman , J. K., Anders, D. E., Fouch, T. D., and Nichols, D. J., Depositional environments, diagenesis,and hydrocarbon potential in nonmarine Upper Cretaceous and lower Tertiary rocks, eastern Uinta Basin, Utah: in Geology of Tight Gas Reservoirs, American Association of Petroleum Geologists Memoir, 35 mscpt. pages.

Ritzma, H. R., 1979, Oil-impregnated rock deposits of Utah: Utah Geological and Mineral Survey Map 47.

Ryder, R. T., Fouch, T. D., and Elison, J. H., 1976, Early Tertiary lacustrine sedimentation in the western Uinta Basin, Utah: Geological Society of America Bulletin, v. 87, p. 496-512.

Tissot , B., Deroo, G., and Hood, A., 1978, Geochemical study of the Uinta Basin: formation of petroleum from the Green River Formation: Geochimica et Cosmochimica Acta. Pergammon Press, London, England, v. 42, p. 1469-1485.

CHAPTER 11

HYDROCARBONS IN FLUVIAL DEPOSITS OF THE GULF COAST REGION

William E. Galloway

The following pages review examples of petroleum resources in fluvial deposits of the Gulf Coast Tertiary Basin. The fluvial depositional architecture of the Gulf Coast was described in detail by Galloway (1981).

FLUVIAL AND DELTAIC FACIES RELATIONSHIPS AND BASIN SUBSIDENCE

The thickness and relationship of fluvial systems to other depositional facies are dependent on the rate of sedimentation compared to the rate of basin subsidence. In coastal plain sequences containing progradational (offlap) depositional episodes, this interplay determines the distribution of fluvial deposits of both the fluvial system and deltaic plain relative to progradational facies of the delta margin (Fig. G-16). Example A, in which rate of subsidence is high because deposition occurs directly upon oceanic crust, is typical of relationships seen in the Tertiary fill of the Gulf Coast Basin. Example C, in marked contrast, reflects rapid progradation onto stable continental crust. This example, which is typical of the shallow water and stable platforms of Mid-Continent basins, results in extensive fluvial incision and cannibalization of progradational deposits. Example B reflects facies relationships in Cretaceous units of the Rocky Mountain foreland basin and seaway.

The Frio (Oligocene) depositional episode typifies the depositional and structural framework of fluvial systems in the North-

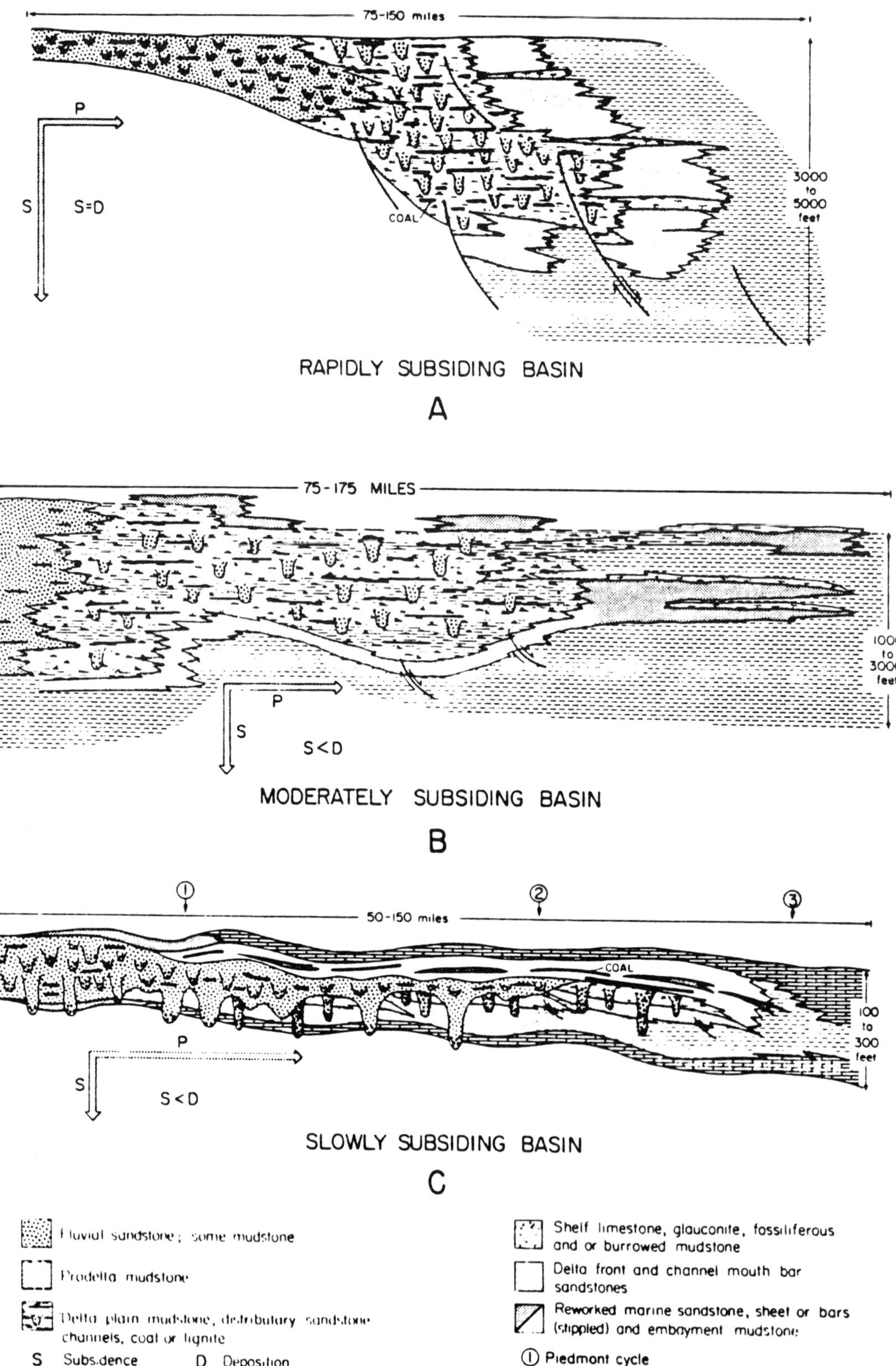

Figure G-16. Relationships between rates of subsidence and geometry of facies and resulting delta system depositional architecture. (From Brown and Fisher, 1980).

west Gulf. Further, it is a highly prolific producer of oil and gas. It will serve as an example for discussion of resource distribution.

FRIO/CATAHOULA FORMATION

Depositional Systems

The Frio Formation (Texas Coastal Plain) and its updip equivalent, the Catahoula Formation (Fig. G-17), consist of several distinct depositional systems (Fig. G-18). Major progradational delta systems, designated the Norias and Houston delta systems, are the principal depositional elements of the late Oligocene coastal plain (Galloway and others, 1982). The Norias delta system was fed by the Gueydan fluvial system, a single, large, extrabasinal, bed-load river system. The Chita-Corrigan fluvial system, in East Texas, fed the Houston delta lobes, and consisted of several major rivers carrying a mixed load of sand, silt, and clay.

Separating the two fluvial/deltaic facies tracts was an area spanning the San Marcos Arch that was traversed by numerous small intrabasinal streams and "creeks." This streamplain system was characterized by deposition of silt and mud between widely spaced, minor belts of locally sourced, coarse-grained fluvial channel fill. Downdip, shorezone deposits record deposition of a large interdeltaic barrier/strandplain system. The massive barrier and strandplain sands were derived by longshore reworking of deltaic deposits and of sediment brought in by small streams of the streamplain.

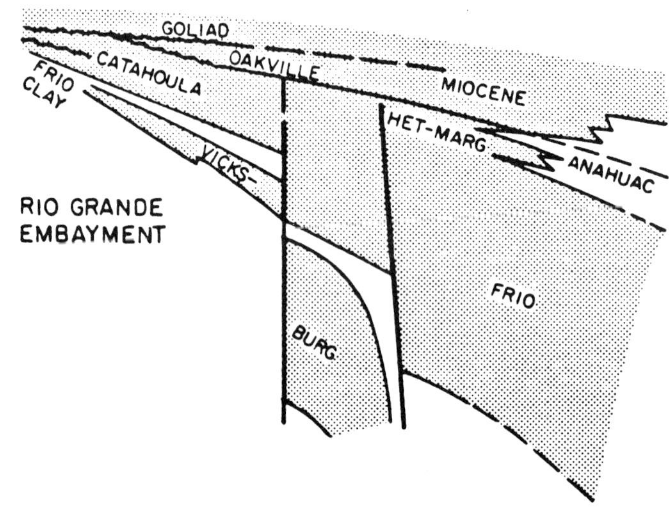

Figure G-17. Stratigraphic relationships of the Frio Formation. (From Galloway and others, 1982).

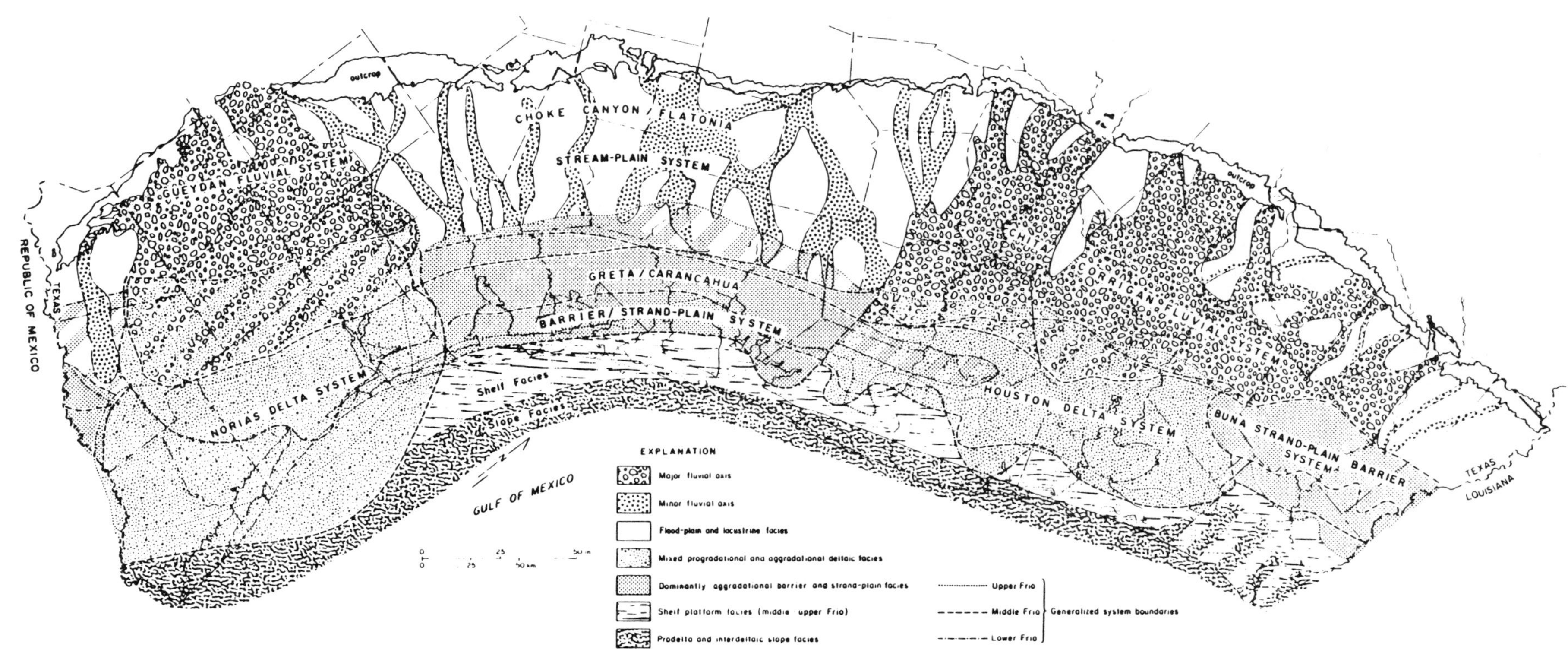

Figure G-18. Frio depositional systems, Texas Gulf Coastal Plain. (From Galloway and others, 1982).

Paleogeography

The paleogeographic distribution of environments reconstructed for the Frio/Catahoula Formations (Fig. G-19) bears a striking resemblance to the late Pleistocene and Modern coastal plain and shelf. Tectonic framework, climatic zonation, drainage element distribution, and coastal and nearshore hydrologic regimes were closely comparable to the present. A primary difference is that early- to mid-Tertiary sea levels were not subject to the short-term glacio-eustatic oscillations characteristic of the Quaternary. These sea-level changes were faster than regional subsidence rates except near the shelf edge; conversely, local basinal factors were more important during Frio deposition. The same depositional phases--aggradational, progradational, and transgressive--are nevertheless represented in the Frio and Quaternary successions, marking a similar response to shifting shorelines and depocenters, albeit on a different time scale. This time factor has resulted in major differences in the thickness of units occupying the same paleogeographic setting.

The late Pleistocene Beaumont Formation, which was deposited under conditions practically identical to those of the Modern, is a near-perfect analog for the Frio Formation.

Hydrocarbon Production

Oil and gas production in the Frio can be grouped into 10 geologically defined plays (Galloway and others, 1982). Principal defining parameters of plays include the depositional system or facies assemblage of the reservoirs, dominant structural style, and type of hydrocarbon produced. Comparison of hydrocarbon yield of the

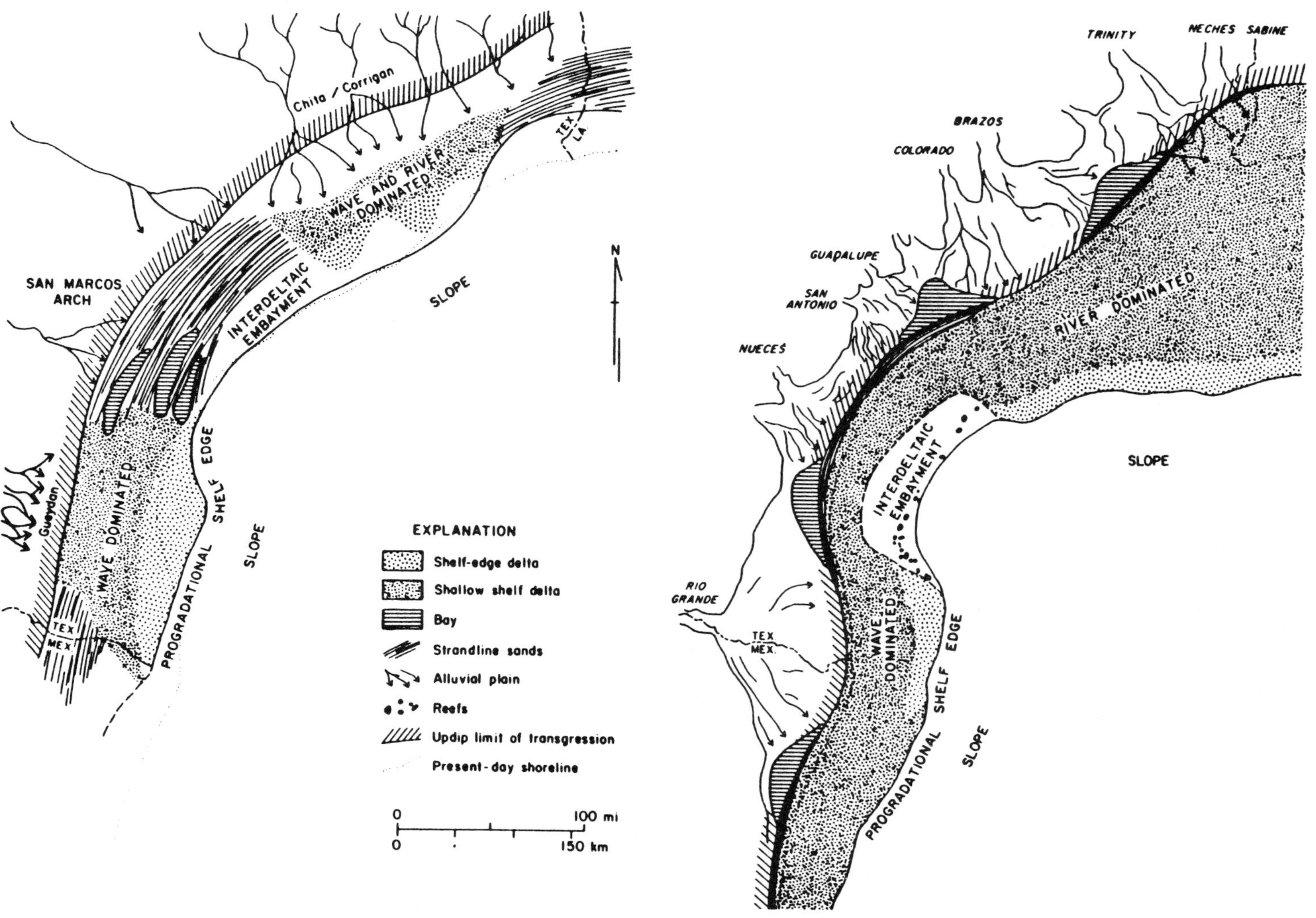

Figure G-19. Comparative Frio (Oligocene) and Quaternary paleogeography of the Texas Coastal Plain and shelf. (From Galloway and others, 1982).

major depositional systems shows that facies are only one factor determining relative exploration potential (Fig. G-20).

The Gueydan Fluvial System

The Gueydan fluvial system is of particular interest because it hosts an anomalously prolific oil and gas play (Fig. G-21) as well as a major portion of the uranium reserves of the Gulf Coastal Plain.

Mapped Extent: The main axis of the fluvial system extends 75 mi (120 km) along strike and 50 to 70 mi (80 to 110 km) downdip (Fig. G-22). In the middle and upper Frio, the system extends across the Vicksburg fault zone. Oil and gas are produced from a series of large rollover anticlines along this structural trend (Fig. G-23A).

Facies: Principal facies include bed- and mixed-load channel-fill sand and conglomeratic sand, laterally associated crevasse-splay sand, and floodplain mud and siltstone. Channel-fill sequences are 10 to 30 ft (3 to 9 m) thick, and commonly stack, producing sand bodies 50 to 100 ft (15 to 30 m) thick (Fig. G-24). Sands amalgamate laterally, forming sand-rich belts up to several miles in width. Log patterns are blocky. Overbank mudstones are commonly highly oxidized red beds.

Petroleum Reservoirs: Hydrocarbons are produced from multiple channel-fill and splay sands encased in the more mud-rich floodplain sequences (Fig. G-24). The large low-relief rollover anticlines that trap hydrocarbons far exceed the scale of the reservoirs; thus production from individual sand bodies is commonly modified by facies geometry (Fig. G-23B). Reservoirs are compartmentalized by channel plugs, which reflect the sinuous channel geometry (Fig. G-25).

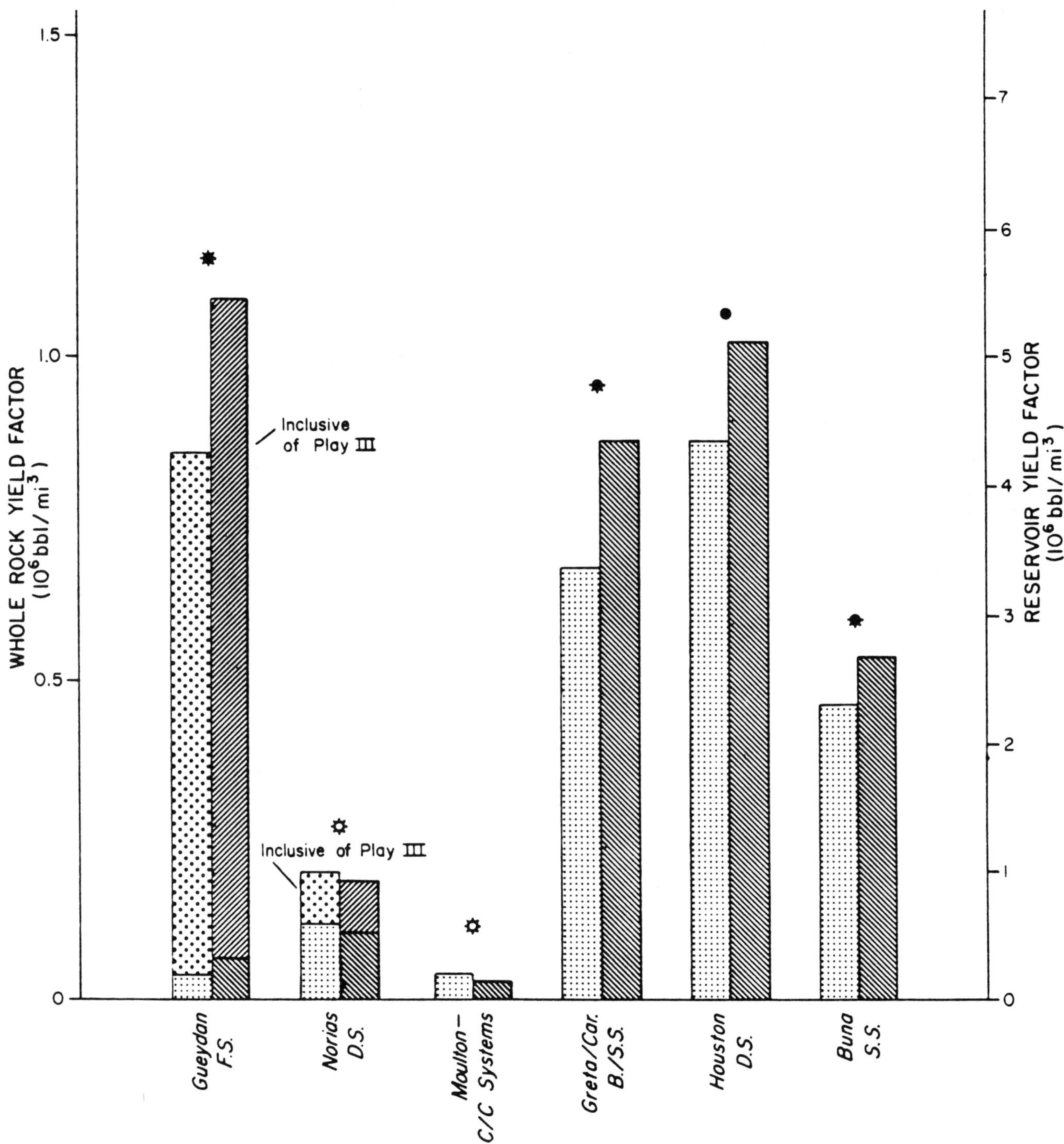

Figure G-20. Frio depositional system productivity. (From Galloway and others, 1982).

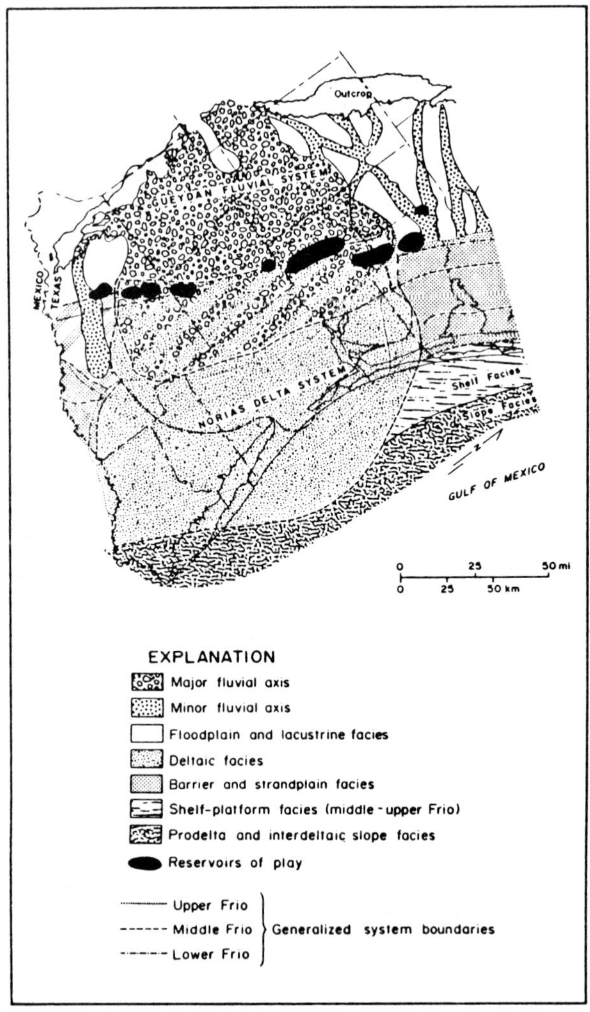

Figure G-21. Depositional systems of the Frio Formation. The fields of the Frio fluvial/deltaic sandstone (Vicksburg fault zone) play lie within the fluvial to upper delta-plain transition of the Gueydan and Norias systems. (From Galloway and others, 1983).

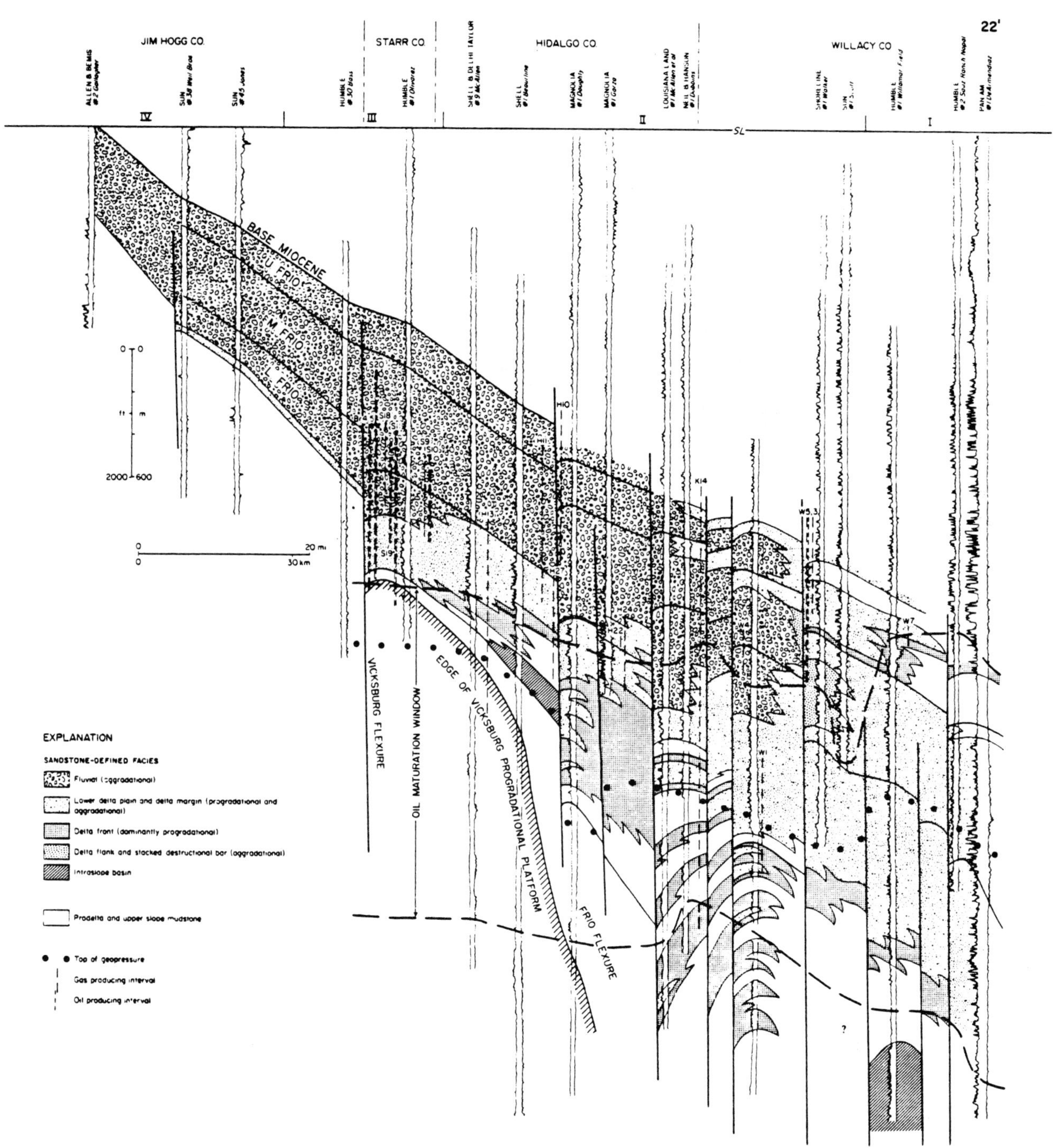

Figure G-22. Dip section along the axis of the Gueydan fluvial system. (From Galloway and others, 1982).

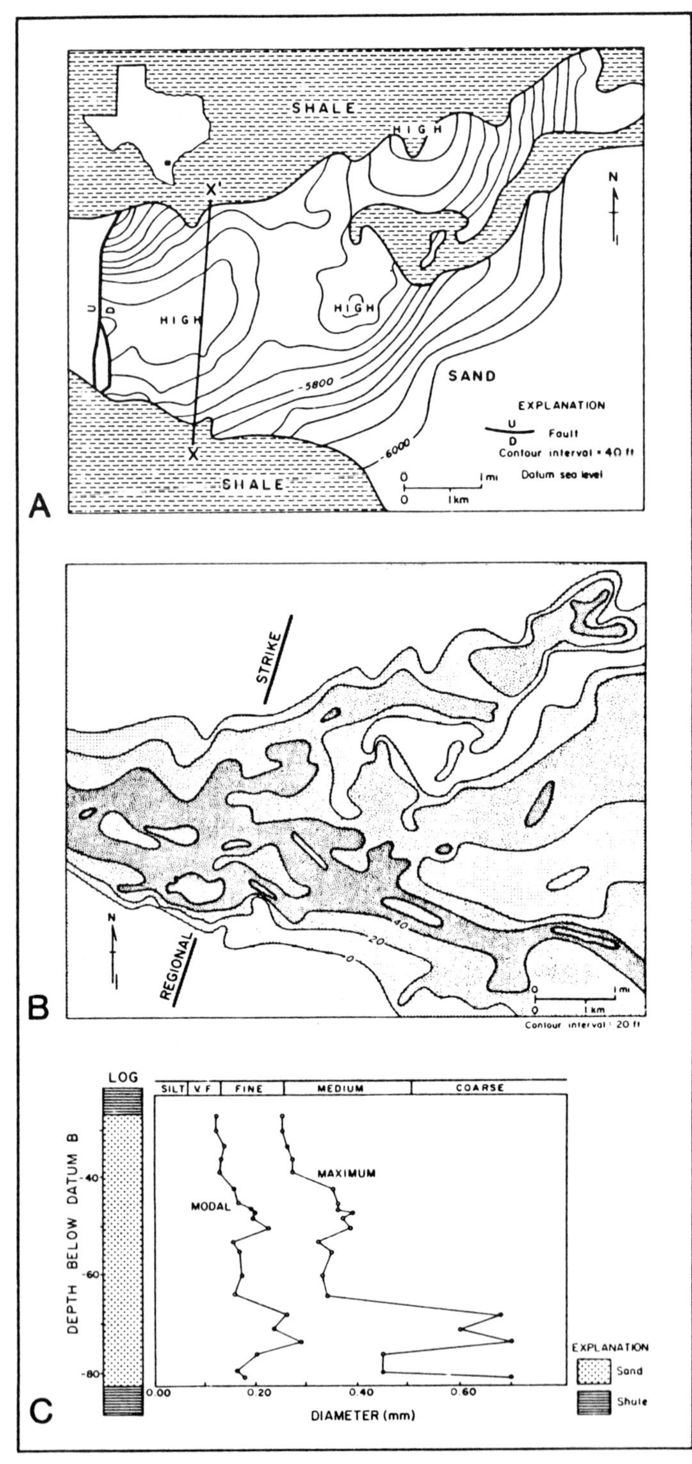

Figure G-23. Frio reservoirs, Seeligson field. A. Structure map contoured on top of 19-B sand. B. Thickness map of 19-B sand. The sand is interpreted to be a fluvial (west) or delta-plain distributary channel fill (east). C. The 19-B sand fines upward, characteristic of aggradational fluvial deposits. (From Galloway and others, 1983; originally in Nanz, 1954).

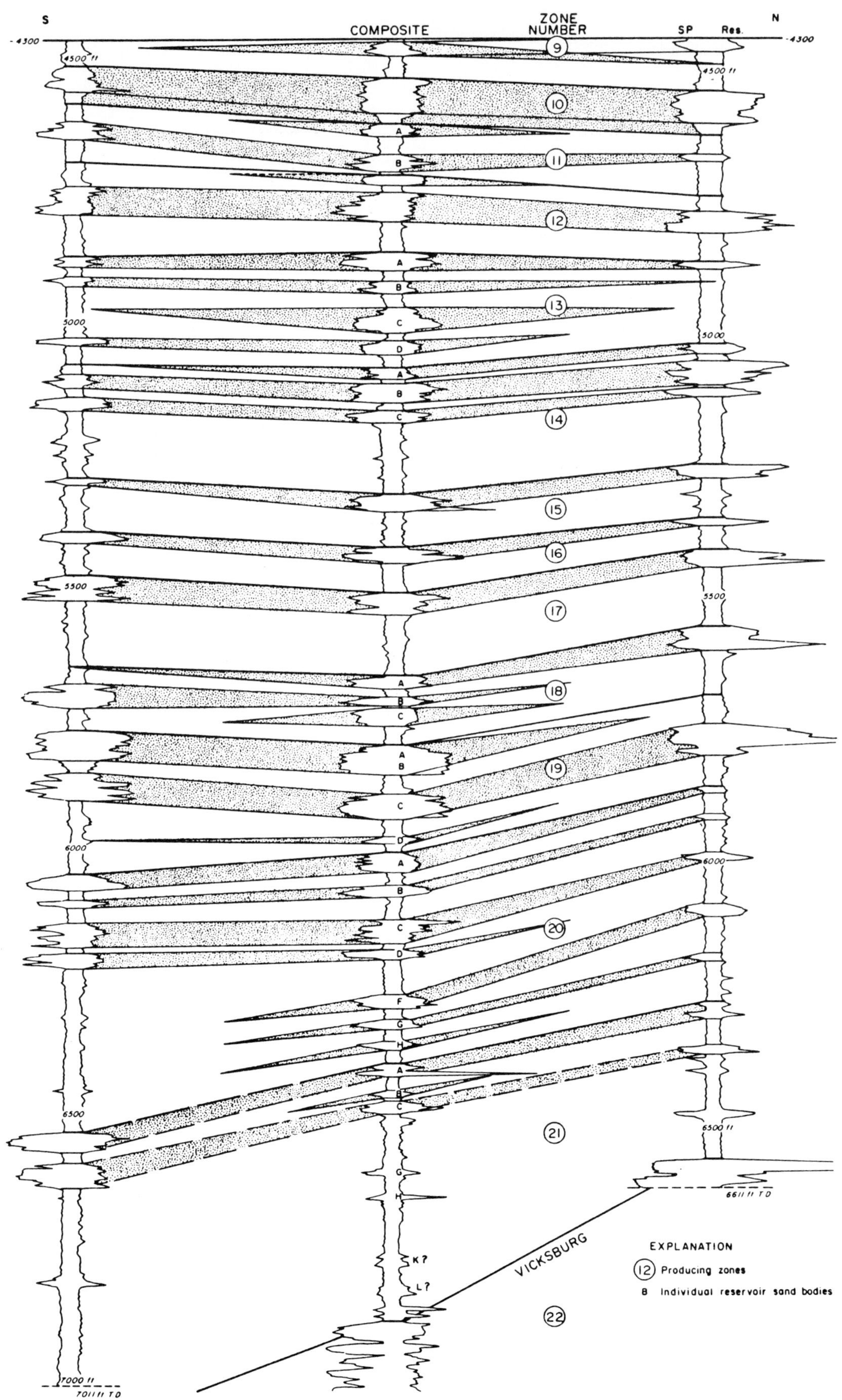

Figure G-24. Type log and correlation of multiple-reservoir Frio sand bodies, Seeligson field. (From Galloway and others, 1983).

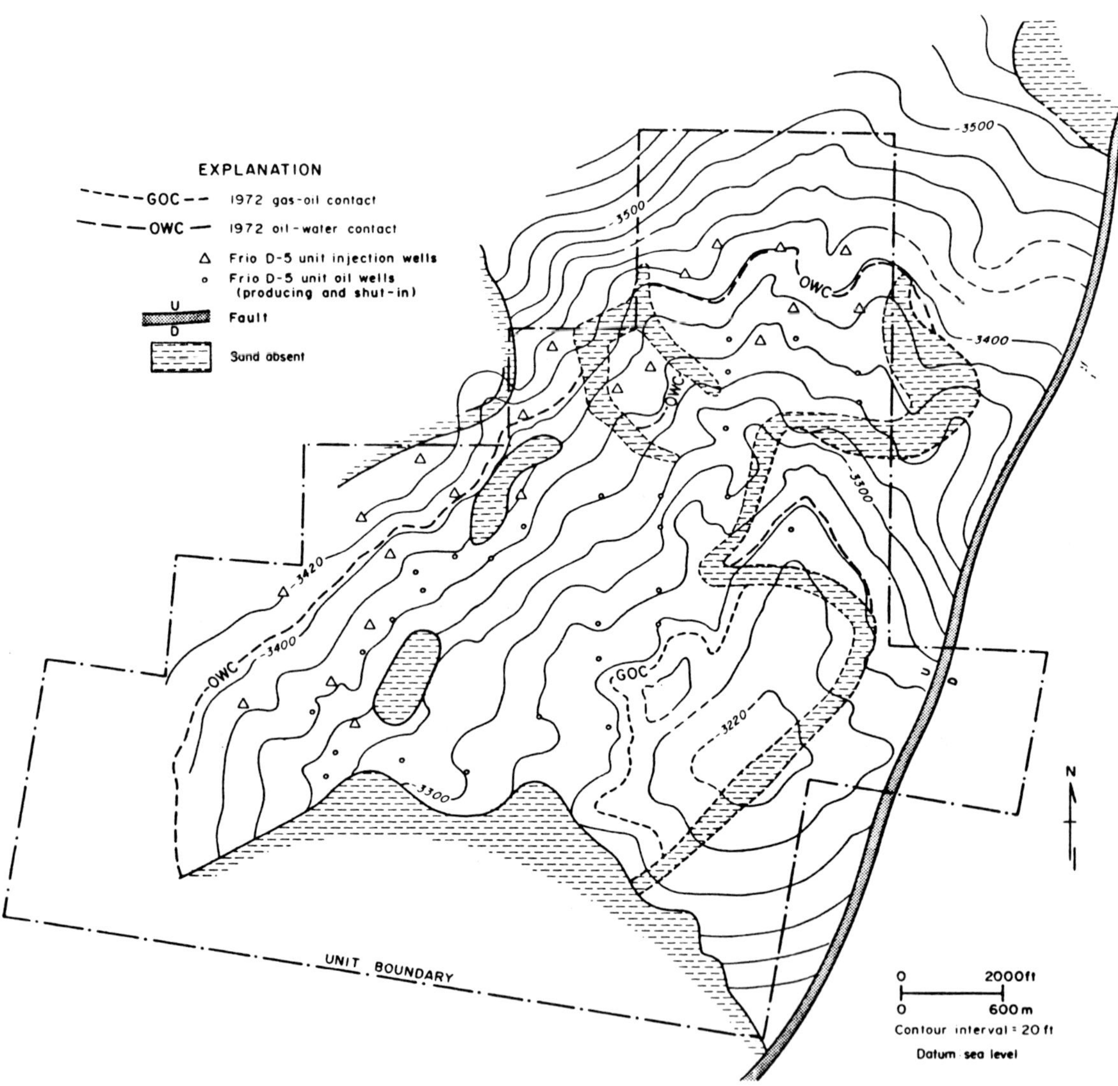

Figure G-25. Structure contour and porosity distribution of the Frio D-5 reservoir at Rincon field. Sinuous, narrow shale-outs, probably mud plugs, cut across the reservoir, creating several intrareservoir pockets of easily bypassed oil and gas. (From Galloway and others, 1983).

Gulf Coast Fluvial Reservoirs -- Summary

Frio fluvial systems typify those of other Cenozoic units of the Gulf Coastal Plain.

1. Fluvial sand bodies are generally not major oil- or gas-producing reservoirs except where they are components of delta systems.

2. Limits to hydrocarbon productivity include the shallow depths of burial, poor quality, gas-prone, thermally immature indigenous source rock facies, and pervasive flushing by meteoric ground water.

3. Significant producing plays are genetically associated with major deep-seated structural features, such as salt domes and growth faults, that tap deeply buried, thermally mature marine facies of older depositional episodes. Large-scale vertical migration of hydrocarbons along structural conduits is characteristic of the Gulf Coast Basin.

4. Traps are commonly structural, but are modified to varying degrees by the complexities of fluvial sand-body geometries.

REFERENCES

ARCHE, ALFREDO, 1983, Coarse-grained meander lobe deposits in the Jamara River, Madrid, Spain, in Collinson, J. D., and Lewin, J., eds., Modern and ancient fluvial systems: Intl. Assoc. Sedimentologists, Spec. Pub. No. 6, p. 313-321.

BERNARD, H. A., MAJOR, C. F., PARROT, B. S., AND LEBLANC, R. J., SR., 1970, Recent sediments of southeast Texas: Univ. of Texas at Austin, Bur. Econ. Geol. Guidebook 11, 120 p.

BROWN, L. F., JR., CLEAVES, A. W., II, AND ERXLEBEN, A. W., 1973, Pennsylvanian depositional systems in north-central Texas: Univ. of Texas at Austin, Bur. Econ. Geol. Guidebook 14, 122 p.

BROWN, L. F., JR., AND FISHER, W. L., 1980, Seismic stratigraphic interpretation and petroleum exploration: Amer. Assoc. Petroleum Geologists Continuing Education Course Note Series No. 16.

CASEY, J. M., 1980, Depositional systems and basin evolution of the Late Paleozoic Taos Trough, northern New Mexico: Univ. of Texas at Austin, Ph.D. Disser., 236 p.

CASEY, J. M., AND SCOTT, A. J., 1979, Pennsylvanian coarse-grained fan deltas associated with the Uncompahgre Uplift, Talpa, New Mexico: New Mexico Geol. Soc. Guidebook, 30th Field Conf., p. 211-218.

COLEMAN, J. M., 1969, Brahmaputra River: channel processes and sedimentation: Sediment. Geol., v. 3, p. 131-239.

FISK, H. N., 1944, Geological investigation of the alluvial valley of the lower Mississippi River: Vicksburg, Miss. River Comm., 78 p.

__________ 1947, Fine-grained alluvial deposits and their effects on Mississippi River activity: Vicksburg, Miss. River Comm., 82 p.

__________ 1952, Geological investigation of the Atchafalaya Basin and the problem of Mississippi River diversion: Vicksburg, Miss. River Comm., 145 p.

FRAZIER, D. E., 1967, Recent deltaic deposits of the Mississippi River, their development and chronology: Gulf Coast Assoc. Geol. Soc. Trans., v. 17, p. 287-315.

GALLOWAY, W. E., 1977, Catahoula Formation of the Texas Coastal Plain: Depositional systems, composition, structural development, groundwater flow history, and uranium distribution: Univ. of Texas at Austin, Bur. Econ. Geol. R. I. No. 87, 59 p.

__________ 1980, Deposition and early hydrologic evolution of Westwater Canyon wet alluvial-fan system: New Mex. Bur. of Mines and Mineral Resources Mem. No. 38, p. 59-68.

_________ 1981, Depositional architecture of Cenozoic Gulf Coastal Plain fluvial systems, in Ethridge, F. G., and Flores, R. M., eds., Recent and ancient nonmarine depositional environments: models for exploration: Soc. Econ. Paleontologists Mineralogists Spec. Pub. No. 31, p. 127-155.

GALLOWAY, W. E., EWING, T. E., GARRETT, C. M., TYLER, NOEL, AND BEBOUT, D. G., 1983, Atlas of Major Texas oil reservoirs: Univ. of Texas at Austin, Bur. Econ. Geol. Spec. Pub. 139 p.

GALLOWAY, W. E., AND HOBDAY, D. K., 1983, Terrigenous clastic depositional systems--applications to petroleum, coal, and uranium exploration: New York, Springer-Verlag, 416 p.

GALLOWAY, W. E., HOBDAY, D. K., AND MAGARA, K., 1982, Frio Formation of the Texas Gulf Coast Basin--depositional systems, structural framework, and hydrocarbon origin, migration, distribution and exploration potential: Univ. of Texas at Austin, Bur. Econ. Geol. R. I. No. 122, 78 p.

GALLOWAY, W. E., KREITLER, C. W., AND MCGOWEN, J. H., 1979, Depositional and ground-water flow systems in the exploration for uranium: Univ. of Texas at Austin, Bur. Econ. Geol. Spec. Pub. 267 p.

HOBDAY, D. K., AND JONES, B. G., 1982, Clastic depositional systems: models for exploration: Univ. of Sydney, Earth Resources Found., 105 p.

JACKSON, R. G., II, 1975, Velocity-bedform-texture patterns of meander bends in the lower Wabash River of Illinois and Indiana: Geol. Soc. Amer. Bull., v. 86, p. 1511-1522.

LEVY, R. A., 1978, Bed-form distribution and internal stratification of coarse-grained point bars, Upper Congaree River, S. C., in Miall, A. D., ed., Fluvial sedimentology: Can. Soc. Petroleum Geologists, Mem. 5, p. 105-127.

MCGOWEN, J. H., AND GARNER, L. E., 1970, Physiographic features and sedimentation types of coarse-grained point bars: modern and ancient examples: Sedimentology, v. 14, p. 77-111.

NANZ, R. H., JR., 1954, Genesis of Oligocene sandstone reservoir, Seeligson field, Jim Wells and Kleberg Counties, Texas: Amer. Assoc. Petroleum Geologists Bull., v. 38, p. 96-117.

SCHUMM, S. A., 1960, The effect of sediment type on the shape and stratification of some modern fluvial deposits: Amer. Jour. Sci., v. 258, p. 177-184.

_________ 1972, Fluvial paleochannels, in Rigby, J. K., and Hamblin, W. K., eds., Recognition of ancient sedimentary environments: Soc. Econ. Paleontologists and Mineralogists Spec. Pub. No. 16, p. 98-107.

__________ 1977, The fluvial system: New York, Wiley, 338 p.

SENI, S. J., 1980, Sand-body geometry and depositional systems, Ogallala Formation, Texas: Univ. of Texas at Austin, Bur. Econ. Geol. R. I. No. 105, 36 p.

SMITH, N. D., 1970, The braided stream depositional environment: comparison of the Platte River with some Silurian clastic rocks, north-central Appalachians: Geol. Soc. Amer. Bull., v. 81, p. 2993-3014.